Articulate Intervention

Articulate Intervention

Hylton Boothroyd

ORASA text No. 1

Taylor & Francis · London

Halsted Press
a division of John Wiley & Sons Inc.
New York · Toronto

1978

First published 1978 by Taylor & Francis Ltd, London
and Halsted Press (a division of John Wiley & Sons Inc.), New York

Taylor & Francis ISBN 0 85066 171 4

Printed and bound in the United Kingdom by
Taylor & Francis (Printers) Ltd
Rankine Road, Basingstoke, Hampshire RG24 0PR

Library of Congress Cataloging in Publication Data

Boothroyd, Hylton
Articulate intervention

(ORASA text; no. 1)
"A Halsted Press book"
1. Operations research. 2. System analysis.
I. Title. II. Series: Orasa text; no. 1.
T57.6.B67 001.4'24 78–13026
ISBN 0–470–26536–1

Contents

Introduction

There can be few subjects which are so misrepresented by their scientific publications as are Operational Research and Applied Systems Analysis. If you read almost any issue of the standard journals, you might reasonably come to the conclusion that it is a branch of mathematics (not necessarily even *applied* mathematics). At the same time, if you read any Newsletter of the OR Society, or any copy of *Interfaces*, a joint publication of the Operational Research Society of America and the Institute of Management Science, you are almost certain to find at least one contribution from an irate practitioner complaining that the journals miss the realities (and the true intellectual excitement) of the subject. This is not a new phenomenon: it has been the same for many years. Why?

The main reason is that we still lack a sound methodological, or philosophical base to the subject. Many authors have over the years tried to explain the nature of the OR process with no more than partial success. Consequently when Boothroyd's first paper on the subject was privately circulated, many of those who persevered to read it carefully experienced what was almost a physical shock. They felt that it is the nearest anyone had got to a description of what really happens. It helped of course that he had been a successful practitioner of the subject for many years, before he moved back to university, but there are whole departments to show that such practical realities can only too easily be forgotten. Since that time, Boothroyd has clarified his thinking and deepened his philosophical basis. The argument, though complex still, travels relentlessly forward.

When a draft of this final version was reviewed by those concerned with the publication of the OR Society, they felt that it *must* be published, but that no suitable vehicle existed. It was the start of the thinking that led to this series being instituted. It is thus fitting that this volume should be No. 1. Many volumes in the series will require less concentration in their reading—few will be as rewarding to those who give it that attention.

JANUARY 1978

ROLFE TOMLINSON
General Editor

Preface

This book is about what we can and cannot mean by the idea of intelligently directing human affairs, about the place of articulating our thoughts before acting, about the place of consultants in business, industry and government, and about the place of formal models.

The book has a lot of theory in it, not because I like theorizing but because sound theory is practical. If you are a consultant, a manager, a lawyer, an economist, an engineer, a scientist, or a mathematician, you will find some of the key issues of your professional world discussed from the theoretical point of view which runs through the book, and as you read the book you are likely to find yourself thinking about and unlocking other key problems for yourself. Indeed, I wrote the book to provide for me and my professional world what I suspect it can also provide for you and yours: a coherent point of view from which to regard high-level advisory, managerial, and administrative activity in an age where naked power is too evident to permit a cosy view of human affairs.

The ideas in the book have come from much wrestling with their validity as statements about my own professional world: operational research. I am conscious of having included as key ideas some of the key ideas of Popper, Lakatos, and Oakeshott. And I am equally conscious of having parted company from them on other matters because of the need to focus on the particularly tricky problem of writing for people at the interface of science, mathematics, consulting, and administration.

Much of the book proceeds by interplay between the general theory and the specific problems of the operational research world. From time to time there are illustrations of a far more general kind: for example, how economies and legal systems might properly be described, the theoretical foundations of statistical inference, the status of ethics, the place of vague language and the unwelcomeness of clarity, the intrinsic incompleteness of science as a basis for action, and the nature of professional competence.

About half the ideas in the book became known to the OR community in the UK through a University of Warwick research paper; an outline of them was presented at the triennial international OR conference in Japan in 1975, and later that year came the proposal that an expanded form of the research paper should be published by the Operational Research Society for a wider readership.

The expansion has taken two main forms. Firstly, an introductory chapter to try to convey to you the feel of my professional world—underneath the specifics, you will find it strangely like your own. And secondly, a lot of new theory which has, I think, removed some of the more awkward jumps in the research paper, as well as extending its application.

There are perhaps two main reasons why what was started for the OR community has a wider relevance.

Firstly, OR is an example of a much more general category of activity: the provision of would-be-expert advice based on knowledge or investigation or both. Many of the problems of describing OR adequately turned out to be special cases of describing the status of the giving of expert advice, particularly that which claims some sort of scientific foundation. The spirit of open-minded scientific enquiry can hardly be said to be the dominant spirit of the present age; would-be-scientific advice often relates uneasily, awkwardly, and naively to the political context within which it is proffered. Moreover, the status of would-be-scientific advice has in some arenas declined sharply as inflated expectations have proved to be ill-founded. This book is in part a reworking of the logical status of expert advice.

Secondly, OR communities in the UK and in other countries address themselves almost entirely to the analysis of other people's decision problems—mostly by invitation, although the invitation is quite likely to have been solicited! A substantial proportion of OR people, particularly the more experienced, see themselves as being prepared to tackle the analysis of any major decision problem in our organized industrial societies *in a spirit of scientific enquiry*. If they do not expect to resolve it completely, they do at least expect to provide a greater insight into the structure of the decision than would otherwise be available. It is a self-concept which may seem unjustifiably ambitious. Yet there is sufficient truth in it for the OR community in some countries to be at present participating in the analysis of decisions which in several ways affect the life of everyone in the country and in many more ways affect the lives of substantial groups within the population. And OR groups in multi-national companies will from time to time carry out analyses which affect the balance of industrial and commercial activities between countries. So what the OR communities busy themselves with is partly your affair! This book is in part about the extent to which they can be justified in their belief that they can make a positive contribution.

I have drawn freely from a variety of writings, but there is no sense in which I started by trying to *apply* other people's ideas. At each stage I have taken only what seemed to contribute to providing solutions to the problems I was dealing with; I apologize if that leads in places to unintended misrepresentation—I am indebted to the writings of several people whom I do not otherwise know, and I have no wish to pain them or those who knew them. I am also indebted to the many members of the OR community in the UK, to Gwyn Bevan and other past and present members of the University of Warwick, and to my family, for a wide variety of insights and specific criticisms. The interventions of Donald Hicks and the generosity with which the OR group at Warwick freed me for a time from teaching commitments were critical in allowing me to complete this first tour of some of the issues of articulate intervention.

MAY 1977 HYLTON BOOTHROYD

1. The O.R. experience

Whilst participation in operational research provides experience which is not easily replaceable by words, there are various ways in which you can get some idea of what the OR community does and how they view what they are doing. Three approaches are provided in this chapter as a background to the theorizing of later chapters: first some examples, secondly a dip into the literature and activities of the Operational Research Society, and finally an indication of some sources where you can read more.

1.1 Some examples

Example 1

The phone rang. It was the university's resident architect. 'I think we have a problem that might be up your street,' he said. He went on to talk about the provision of lifts in a new building that was still at the design stage. Money was tight. In any case there were national norms for the provision of lifts in university buildings. But what would happen when the building was fully in use in a few years' time? Would standard lifts be adequate? Would alternative larger lifts be adequate if standard lifts were not?

I knew little about lifts. But I *did* know about the idea of imagining how a system will work before all its component parts exist or are brought together—of simulating how events will succeed one another through time so that unanticipated operating awkwardnesses are known beforehand and can be prepared for or avoided altogether by redesign. So even before our phone conversation was over I had a fairly clear idea of the information we would need and how I would use it.

From the likely supplier, one of the architect's staff got data on technical characteristics which he and I soon reduced to:

> Time-per-floor for non-stop running,
> A single decking-time to include the extra time for deceleration, doors opening, waiting with doors open, doors closing, and acceleration.

The times were close to those for lifts in existing buildings. We did not know how long people would take to enter and leave the lifts—a visit to the university library at a busy time of day soon gave us a relationship between the number of people passing through lift doors and the extra time the doors were held open. So by that point we had a working understanding of how the proposed lifts would operate, but no model of the social environment to which they would be expected to respond.

With the person who timetabled university activities we looked at the incidence of group teaching and came to the conclusion that the busiest changeover periods would be around 11 a.m. and 3 p.m. each day. From projected numbers of staff and students for the building we estimated how many people would be entering and leaving the upper floors of the building during those busy periods, ignoring those few who might make intermediate trips between floors.

But how would all these future people behave? We took what we thought were slightly conservative views that favoured the smaller lifts: no-one would use the lifts for the ground, first, and second floors; only half the third floor people would try the lifts; all the fourth and fifth floor people would use the lifts; people would arrive in ones, twos and threes over a 10 minute period; and tutorial groups of various sizes would finish at times that were also scattered over a 10 minute period. These views were consistent with the university as we knew it, and with the ways in which we noted people behaved now that we had a reason for paying attention to particular aspects of their behaviour.

It was next a simple technical matter for me to construct a few samples of future changeover periods, using a method which scattered arrivals and tutorial-finishes about the busy period in such a way as to mimic the lulls and short bursts of activity that characterize a busy organization, rather than supposing that everything would happen at nicely spaced intervals. With these schedules of wished-for journeys I could then play through on paper how the different possible lifts would cope. In my simulations on paper, the standard lifts became inadequate well before the end of the busy period; the larger lifts remained just adequate.

The simulations gave me some insights I had not expected, obvious enough once I had seen them, but unlikely to have occurred to me by pure thought:

(1) As traffic levels build up, a lift which has been coping quite well visiting such floors as are necessary suddenly gets slowed down by a rapidly self-aggravating process so that it is pushed into a maximum-length cycle in which every floor is visited in response to calls from would-be passengers, but would-be descenders on the third and fourth floors are confronted by an already full lift—as traffic builds up the lift therefore lurches into a slower cycle with non-productive visits and then is effectively locked into that cycle,

(2) The effective carrying capacity of the lift under heavy traffic could be kept high by designing a load-sensing device which when activated by a full load would cause calls to be ignored in favour of the passengers on board—non-productive stops would be avoided and a quicker cycle would be maintained.

In all, it only took two or three days from starting the problem to presenting a two-page report, one page of which was a diagrammatic representation of a busy period. The larger lifts *were* chosen. At peak times they *are* busy; as I pass the waiting group on the way to my office I know that no-one there is aware of what daily frustration there might have been!

I have no continuing interest in lifts. I chose this small example because it is within the experience of many people, and simple enough for the mix of social and technical content to be accessible to a varied audience. It is a typical example in its interplay between people and in its interplay between precise statements and imprecise understandings. But it was about an untypically minor matter, with untypically simple technical content, posing untypically simple problems of understanding, measurement and evaluation, and it was therefore untypically quick.

The idea of simulation has been very widely used. It would be difficult to think of an activity which has not at some time or other been simulated either with responsible intent or for fun as part of an educational programme. There are several instances of whole ports being simulated either to help in specifying new facilities or to learn how to operate existing facilities more effectively. Likewise, we can find reports of simulations of mining operations, fire services, steelworks, computer operations, communications systems, maintenance activities, military conflicts, supply systems, production control, financing activities . . . the list is almost endless, and is being added to every week somewhere in the world. The building of a major simulation model can take months, and can be part of a sequence of investigations taking a few years, though that is only warranted where the end product is to be a method of rapidly constructing and operating simulation models for a class of problems which is known to be recurring and known to be significant. For example, several man-years' effort went into the construction of computer programs whereby mining engineers in the UK can rapidly construct and run models of new mining operations where the balancing of underground storage and transport can be critical for uninterrupted high output; much of the research effort went into understanding how the initial elaborately realistic but time-consuming models could be slimmed down without losing their validity, and into making the models rapidly alterable so that they could keep pace with the planner as his insights developed in the course of analysing a particular mine layout.

Not surprisingly, such a popular approach is supported by a wide-ranging abstract technology—a technology of analysing what

Ackoff and Emery called abstract systems—systems all of whose elements are mental constructs. By the time I reached my simulation on paper in the lift example, I was dealing with an entirely abstract system and representing aspects of it on paper. (Do not be misled by the fact that I *hoped* the abstract system would be like a future real system! All the elements of my model were constructs which I had specified, and which I was free to change as I chose. Indeed I was free to choose to specify an abstract model which I believed to differ from what the real future might be.)

The earliest elements of the abstract technology for simulation were ideas already existing in the world of statistics—ideas of probability mechanism and ways of appraising the large quantities of numerical information that simulations could produce. It was common for an OR team to translate all these ideas into a computer program from first principles. Later, libraries of standard sub-programmes became available, although there were so many libraries in so many different computer languages that this was hardly the blessing it might have been!

More importantly, a few people pondered on the structural similarity behind the physical differences between lifts, and ports, and maternity wards, and aircraft engine maintenance systems, and the like. In each case one type of *entity* (e.g. a ship) would arrive and *wait* for other types of entity (e.g. dockers, a crane, railway wagons) so that they could together go through a common *activity* (e.g. unloading the forward hold) and would then be free to go on their separate ways for their next rendezvous with other entities. Here was the beginning of a set of abstract constructs which could provide a formal language for simulation. Provided you can match the elements of a real system to the elements of an abstract language, you can in return have all the facilities that the new language offers. Since it is a formal language, someone will typically have already arranged for sets of statements in the language to be automatically translatable into a computer program for playing-through the operation of any system describable in that language and for reporting on what happens in the simulated operations.

Within the worldwide OR community or just across its boundary in the world of computing, there are by now several dozen people who have written or shared in writing computer realizations of abstract simulation languages. The variety of languages is at times confusing to the rest of the OR community, but it does free them to concentrate on the central matter of modelling each particular real system they are concerned with analysing. The latest computer methods for realizing abstract simulation models provide a helpful range of prompting and automatic indications of logical loose-ends. They make possible in several weeks what would earlier have taken several months. But although the computer methods can ensure that a simulation model is internally self-consistent, there is nothing inherent in them which *ensures* that the finished model corresponds with the reality it is intended to represent. Nor could there be. Tests of correspondence are a matter for people and, even at their most critically imaginative, they can logically go no further than to indicate that the correspondence between model and reality is not yet disproven.

In the OR world, simulation is only one of several major and many minor themes on how to represent our abstract conceptualizations of actual and imagined real systems. All the themes are about taking what might be loose pictorial or verbal conceptualizations and making them more explicit, more precise, more internally elaborated, more complete, more free of irrelevancies and inconsistencies, or more computable. Some of the themes, of which simulation is one, are about the abstract technology of passing from statements of what might be done to statements of what might be the consequences. Other themes are about constructing a single complex representation of many alternative things that might be done in the abstract system together with their many alternative consequences, and of then having an abstract technology for searching for what interests you.

Example 2

Our late Saturday lunch had taken us well into the afternoon. Our original plan to go shopping now looked impossible, since no start had been made on dinner for friends due at 7.30 p.m. and each of the three rather special courses would need quite a lot of preparation.

On impulse, I found pencil and paper and switched into my professional role. Five minutes was enough to elicit from my wife a fairly long list of jobs and cooking operations, how long they would take, and what necessarily had to precede what else. At the back of my mind I had the notion of looking for that sequence of activities that would take longest—the critical sequence—and the notion of building a plan backwards from the endpoint it was intended to achieve. The plan would have to be implementable by the limited labour force available, and so it could potentially contain many alternative sequences of jobs and operations.

The first draft plan was not encouraging. It gave us no more than an hour for shopping, and on a busy pre-Christmas Saturday that would be entirely taken up by travelling. However, a formal sketch of the plan led to two insights:

(1) We had somewhat rigidly excluded the time after 7.30 p.m.—by planning some of the unattended heating, cooling, and setting of later courses to continue to the time they would be needed, and by permitting short finishing operations, we could shift segments of the plan to a later time; we could then expect to follow a leisurely (but undeclared!) schedule of course timings,

(2) There was undeniable scope for shifting simple jobs like vegetable preparation from my wife's part of the plan to mine!

The revised plan had a starting time of 5.30 p.m., a time so late that we had difficulty in believing that the logic of our estimates and calculations could be correct. On the face of it we had, from 10 minutes or so of exploratory calculation, gained an extra hour and a quarter. It looked too good to be true.

We decided to implement the plan. The shopping was a success, and so was the rest of the day. The times all turned out to be close to what had been estimated, there were no surprises from what might otherwise have been overlooked, and there was no need to revise the plan part way through.

The success of that simple planning exercise rested on two things, firstly a reliable imagination based on experience of what would need to be done and how long it would take, and secondly some compact and convenient way of representing and manipulating our abstract conceptualizations of what we took for our purposes to be the essential characteristics of the rest of the day. A later attempt to plan the sequences for decorating a room was a fiasco; my initial picture had several significant omissions and I had little idea how long the various operations would take.

The introduction of a formal planning technology does not of itself ensure success even when the intended operations are well understood:

> Two rather similar reorganizations were to take place in different parts of the UK. Each would lead to the commissioning of new buildings, the disposal of old buildings, and the relocation of much material whose movement would have to be documented in detail. The more complex of the two was completed in 3 weeks more than its initial 50-week schedule; computer methods were regularly used to update and revise the plan. The simpler of the two reorganizations was still incomplete more than a year after its initial target date: computer methods were regularly used to update and revise the plan.

In the first reorganization, the plan was an intrinsic part of what happened. In the second reorganization, the plan served only to emphasize the lack of will to implement what had been decided elsewhere.

There is an extraordinary variety of structure to problems of planning what to do, so much so that the variety of abstract technologies so far constructed for searching through alternatives can deal adequately with only a modest proportion of the abstract systems which have been conceived as representations of real systems. Moreover, apparently quite minor changes to the structure of an abstract system can change it from one for which a compact economically-realizable abstract technology exists into one for which the best that we can do is to construct what we hope are enlightened guesses for searching. So when an OR analyst has conceived an abstract system he hopes corresponds reasonably to the real system, he is quite likely to have a problem of finding or devising a suitable abstract technology, or finding or devising a suitable computer realization of an abstract technology, for exploring properties of the abstract system. The OR analyst may not experience too much difficulty at that stage for both good and bad reasons. He is quite likely to have studied a wide range of abstract technologies as part of a graduate or undergraduate programme, and is likely to have experience of searching for or constructing computer realizations; on the other hand his exposure to abstract technologies might be so high that his thoughts about real systems are limited in significant and distorting ways to just that restricted range of concepts that will lead on to just that restricted range of abstract systems that are conveniently explored by currently available technologies. Indeed, most educational programmes are biased towards the study of abstract technologies and are biased away from the study of educing abstract systems from real systems and of examining the correspondence between abstract and real systems.

The concentration on abstract technologies is understandable: they can be studied in the privacy of your own home and they are readily transmissible between countries since modern industrial societies make the study of elementary forms of abstract technology virtually compulsory (everyone spends many hours doing mathematics!). The lack of attention to eduction and correspondence is also understandable: they require an abstract system, perhaps several competing abstract systems, to be juxtaposed with a slice of life which by its nature is not transmissible and they require a mental stance of critical enquiry which is only weakly present in classrooms —whatever aspirations there may be to the contrary, teachers and students find it mutually convenient to keep the teacher in the role of authoritative provider of knowledge. Indeed, critical enquiry into locally convenient social activity like city administration or manufacturing or banking can arouse extremely strong feelings!

Nevertheless, the OR communities in some countries have a tradition of carrying out, and according high respect to, studies in which the prime emphasis is on drawing out a structured understanding where there was little previous structure, of identifying important factors and making well criticized estimates of how they interact with each other, and on devising meaningful measures of evaluation where what previously existed was unsatisfactory. In such studies, what I have called abstract technology may well be absent or elementary, though it sometimes forms a major later part of a study in its own right. The tradition of carrying out classical studies of educing structure is strongest in countries with the longest experience of OR. Indeed, without classical studies OR communities would not exist as we know them. Many widely repeated themes and many widely recurring topics of study had their origins in what were in their own time classical studies of what had previously been inadequately structured. However, not all classical studies have direct descendants; some have their main value to us as exemplars of imaginative critical enquiry where the detailed understandings that emerged from the enquiry were not usefully transferable to other types of problem or to more than a handful of other contexts.

In wartime Britain, established scientists came from the study of physical and biological problems to the study of operational problems by various routes. Some went on from their prewar study of radar devices to the study of the opportunities that such devices offered for the restructuring of systems of command and control for air defence. Others who were already distinguished in peacetime fields saw as outsiders the potential for scientific approaches. Indeed, one group wrote and had published in a matter of days one of the first instant paperbacks, *Science in War*, as part of their personal initiative to gain an entree to military and supply operations which they had identified as having potential for improvement if studied by people like themselves: people whose qualification was not that they had any special knowledge of the operations but that they were grounded in the observational and argumentative traditions of science. Fortunately they made their point.

Bringing supplies across the Atlantic produced many strategic and tactical problems about which there was little that could be said from previous experience. For example, the use of aircraft to detect and destroy submarines was a novel form of combat in which the aim for British aircraft was to take surfaced submarines by surprise and to sink them by dropping depth charges on them or into a tightly delimited area close to where they had disappeared. The team studying anti-submarine operations kept up a lively interaction between the real system and their growing abstract understanding of it by flying on operational trips and by introducing new forms of recording and reporting information on searches and attacks, some of which conflicted with optimistic crew reports. What they were looking for were not simply ways of giving appropriate structure to their abstract understandings but also ways of quantifying relationships. So, for example, one could argue that flying higher would allow a greater area to be swept in a given time, would bring a greater number of sea surface objects into positions with a direct line of sight to the aircraft, would reduce the chance of noticing any *particular* object in the greater area swept, would reduce the chance of getting close before being noticed, and would affect the chance of a particular attack leading to worthwhile damage or sinking; in order to have a quantified view of the net overall effect of height one therefore needed a quantified understanding of all these separate elements supported from existing operational records or from new operational records or from observational trips and experiments. Moreover, for the understanding of the effect of height to be useful, it would have to be reachable quickly in response to the continual stream of technical and tactical innovation on both sides. In effect the OR team needed to, and did, make itself the centre of a science-driven intelligence service.

The interlinked studies produced several surprises. Some came from looking at existing records in new ways: for example, when sightings were charted to show where they occurred during the standard 30 minute observation period, there was a clear falling off in the number reported after the first 10 minutes—no-one had suspected that there was so rapid but remediable a decline in the performance of observers. Others came from being present during operations: for example, it occurred to one observer watching depth charges exploding round a recently submerged vessel that it would not by that moment have had time to reach its standard operating depth, at which the depth charges had been set to explode—an alteration to the depth setting, which required the design and manufacture of a new device, led to so sharp an increase in power to damage that an intercepted signal was heard to report the introduction of a new type of explosive!

Waddington's account of these and other studies was written just after the war, but at the last moment its publication was deferred by 25 years. Fragments of the studies were known from other sources but these did not fully capture the experience of teasing-out complex interacting understandings, which sometimes ran sharply counter to the prevailing wisdom.

Each new industrial, commercial, or administrative context for OR studies provides a fresh opportunity for classical OR, although the opportunity may be passed over or deferred in favour of simply looking for ways of using existing themes from elsewhere, or in favour of simply formalizing a sponsor's understanding without making any critical appraisal of that understanding. However, over the years there has been a steady stream of classical studies producing models of activities as diverse as the spread of forest fires and the effects of firefighting measures, trawler fishing, policies for operating opencast mines, hospitals and health care, steel-making, banking, and the effectiveness of crime-prevention methods.

1.2 Some evidence from the Operational Research Society

On my shelves as I write is a long sequence of issues of *Operational Research Quarterly*. Each issue carries a page of information about the Operational Research Society, whose journal it is. Among the slowly changing lists of council members, study group secretaries, and chairmen of committees are several colleagues from my early days in OR, many people whom I know less well, and some whom I do not even know by name. It would appear that they jointly concur with the opening statement on what we may take to be their page, since the statement has for some years remained unchanged as:

> Operational Research is the application of the methods of science to complex problems arising in the direction and management of large systems of men, machines, materials, and money in industry, business, government and defence. The distinctive approach is to develop a scientific model of the system, incorporating measurements of factors such as chance and risk, with which to predict and compare the outcomes of alternative decisions, strategies or controls. The purpose is to help management determine its policy and actions scientifically.

An earlier version spoke of OR as the attack of modern science on complex problems, and in doing so was perhaps more firmly in the tradition of expecting that science will of itself dispose of the problems of mankind.

No-one is asked to subscribe to the statement about OR; it is not a touchstone of doctrinal purity! Its overt purpose is to inform whoever may care to browse and perhaps it also serves as a rallying point for those who see themselves as part of the OR community. As such, there are many members of the OR community in the UK who feel it does not *quite* say what they themselves would wish. But although a competition for a replacement had many entries, none of the competing statements got enough votes to replace it. So it remains. It continues to irritate those who do not wish to see OR defined as exclusively for those who currently have managerial responsibility. It continues to haunt those who know that the elaborate models they work with owe more to the concepts of an abstract technology than to scientifically well-founded concepts about the particular real systems they confront. And yet because it contains some sort of idealization of OR it continues to challenge the OR community to think ambitiously, to think critically and to think in terms of classical studies rather than in terms of simply applying ideas from elsewhere.

The facing page carries information about the *Quarterly* itself: who is the editor and who are his advisers, where to write for various purposes, and an opening statement on editorial policy, which has also remained virtually unchanged for some years:

> *Operational Research Quarterly** is published eight times a year on behalf of the Operational Research Society Limited, London, and is addressed primarily to the practitioner of OR. Contributions on any matter relevant to the theory, practice, history or methodology of OR or the affairs of the Society are welcome but it is specifically the aim of the Publications Committee to encourage the submission of accounts of good practical case studies illustrating OR in action; of reviews of the state of development of fields of knowledge relevant to OR and reviews of the use of OR; and of controversial articles on methodology, technique or professional policy.

Publishing the *Quarterly* eight times a year is not whimsical; after many years as a true quarterly, increases in circulation and the volume of worthwhile contributions warranted an increase in the number of pages published each year, and that was more easily achieved by limiting the size of individual issues and having more of them! More importantly, the statement is not simply a statement of what will be preferred from what is offered, but it is also part of a continuing attempt to influence what is written about; for the *Quarterly* has only occasionally had the balance of contributions suggested by the statement of editorial policy.

There are relatively few case histories illustrating OR in action, although in a study made by a group of my final year students the *Quarterly* compared well with other journals. For the study we formulated a set of criteria along the following lines, and scored each criterion subjectively on a five-point scale:

(i) Does the case show evidence of being the author's first-hand experience of producing a structured understanding of a real system and is the experience of drawing out an understanding well reported?
(ii) Are checks of the correspondence between the abstract system and the real system sufficiently extensive and well reported?

* Now entitled *Journal of the Operational Research Society.*

(iii) Does the case illustrate the good use of abstract technology?
(iv) Does the case report well how the abstract technology or its results were to become part of the social process of the real system and were there sufficient predictions of the consequences of introducing the changes?
(v) Does the case report the implementation of what was recommended and does it include comparisons of predictions with what later happened?

Few case histories from any source scored well on all these. Those that did were likely to have won an award from one of the OR communities or at least to be widely cited. Although the boundary between case histories and other forms of paper was hazy, the group had no difficulty in collecting a couple of hundred examples, whose pattern of scoring on the five criteria was typically:

fair/poor/good/quite good/absent.

Given that an OR case history requires a sponsoring organization to approve a degree of self-revelation well beyond what is normally practised, given that practitioners have fewer incentives to publish than academic developers of abstract technology, and given that there is a sharp distinction between being able to do something and being able to describe the doing of it informatively, I found the exercise encouraging.

However, the preponderance of what is offered and published falls under the headings of 'theory' and 'technique'. At first sight that must be good. What could be better than the provision of theory and technique? But it has for many years been the custom in the OR community to limit the range of what is indicated by the word 'theory' to symbolic explorations of the properties of abstract systems and similarly to limit what is indicated by 'technique' to the abstract technology established by way of such explorations. Someone coming freshly to the area might reasonably have expected that the two words between them would have included notions of how abstract systems can be drawn out of real systems, and how abstract technologies are to be used within the social processes to which they are supposedly contributing. Not so. And because the abstract systems devised by OR people hardly ever include an explicit representation of the abstract system being used in the real system, there is little in the study of abstract systems that is likely to shift perceptions outwards to eduction, correspondence, and social process.

Nevertheless, a fair proportion of the papers about abstract systems contain a brief opening account of how the abstract system came to be specified as part of an advisory relationship. It is rare for such accounts to contain any hints of the false starts and critical appraisals that attended the construction of the abstract system: much more commonly the construction is reported as though it had been quite unproblematic in words such as 'We found that . . .' or 'The problem can be formulated as . . .'. When I talk to authors I find some of them considerably more aware of the critical processes of eduction than they have conveyed in writing: others find it almost insuperably difficult to distance themselves sufficiently from their perceptions to appraise the quality of what they have done. My impression from these papers is of people doing a reasonable job of modelling real systems into abstract systems, but being rather inarticulate about it and being only partly self-aware.

The rest of the papers about abstract systems carry no evidence that the author has drawn out the abstract system from some real system or set of real systems that he has entered. True, the words in them may well appear at first glance to be about real systems, since words like 'men' and 'vehicles' and 'patients' and 'routes' are freely used. But they appear to have limited stylized meanings that have carried over from accounts of other abstract systems. Indeed the motivation for considering the abstract system is likely to be that a new analysis of it can be devised or that it is a new variant of an abstract system previously written about. My impression from these papers is mostly of people with a vigorous capability for devising abstract technology, but with weak or untried ability in modelling from real systems and with weak or untried imagination of how their work could be embedded in a real social process.

It is the gap between those with a high capability in abstract technology and those whose strength lies in the diagnosing of what Oakeshott calls the goings-on within societies which is the most serious structural weakness of the OR communities. Conversely, the strength of the OR communities lies in those who combine in themselves the ability to draw out structured understandings of real systems and the ability to devise and adapt for ongoing use whatever abstract technology is needed.

Over the years the *Quarterly* has also carried a variety of contributions on the nature and practice of OR, including reflections by the Society's incoming presidents, surveys of the background and activities of members, and papers intended to provoke and extend. In spirit this book is a contribution to that stream, although I feel that in some ways what I have written distances me from the stream, since I am explicitly or implicitly calling into question much that is in it. There is a magic about people creating new understanding for which words like 'the application of methods of science' suggest something too mechanical and dull. There is an unexpectedness in the development of human society that makes words like 'determine its policy scientifically' appear to belong to too static a concept of the world. There is a potential for the harsh exercise of power in the face of which 'helping management' lacks moral energy. So my aim in what follows is to work through some of the issues of doing OR in ways which will stand up in very different social contexts, which deal equally well with successes and failures, and which are direct enough for practitioners and abstract technologists alike to adopt as a point of view from which to construe their activities.

1.3 Some sources of further reading

If you have access to a reference library, you can learn a good deal about OR in the course of looking for and browsing through case histories in the British journal, *Operational Research Quarterly*, the two American journals, *Operations Research* and *Management Science*, the Canadian journal, *INFOR*, and the proceedings of the international conferences which have taken place at three-yearly intervals since 1957. It is in any case worth reading the book by Waddington and selecting some of the cases from the two volumes whose editing McCloskey shared.

Books about the abstract technologies of OR are legion at all levels of sophistication, since there are many opportunities for them to compete for adoption as course texts. The books by Wagner and by Hillier and Lieberman give an idea of the range of abstract technology which is studied in undergraduate and graduate programmes in the UK.

Perhaps the most useful point of entry to writings on the doing of OR—its methodology say—is the survey by White. In it he sees himself as imposing order on his large collection of source material in the traditions of the writings of Ackoff and Churchman, whose many books and papers have provided much of the conceptual climate and vocabulary for what has been written about the goings-on of OR, and whose ideas I too have freely drawn on whilst constructing a somewhat different viewpoint.

1.4 Some orienting remarks

In this first chapter I have tried to convey something of the experience of doing OR. I have done it in terms which only partly overlap with those in common use in the OR world. That was deliberate. I needed something outside its own present language to describe it in ways which would sharpen what is perhaps only half perceived. In fact, I have already been using the point of view which the remaining chapters are about.

Behind the specifics of doing OR and belonging to the OR community, there is the more general theme of belonging to any professional community which has a large body of generalized understanding which the members seek to relate to particular problems in the real world. The OR community is a particular example of a much wider group which includes economists, engineers, planners, applied scientists, and financial executives as near-neighbours because of the similarity of the important role that abstract technology has for them, and which includes doctors, lawyers, management consultants, social workers, administrators, and architects as more distant neighbours because of the common commitment of focusing general understanding on to particular here-and-now problems of action.

For all professional communities there are underlying darker themes. There is the theme of professionals fully understanding neither themselves nor the status of their advice. And there is the related theme of professionals having their own language which, despite motivating them and ordering their activity, misleads them and leaves them without the linguistic means for prompting the recognition of gaps and misconceptions and for articulating them.

My method of proceeding in the following chapters will be to discuss a linked sequence of ideas which are applicable to any kind of thoughtful behaviour. I shall aim to establish their meaning by appealing to your experience and your sense of human history; the ideas therefore stand in permanent risk of being shown to be false by you or any other reader. I shall frequently 'zoom in on' issues from the world of OR, many but not all of which are issues in the wider world of giving any form of professional advice.

2. Programmes

There is about every ongoing human situation a totality which can be described from many points of view. We could examine the biochemistry of the individuals and wonder at the splendid complexity they represent. We could concern ourselves with their biophysics: with the rich field of mechanical and electrical activity and with the relation of stimulus to response. We could describe the locations of the actors in space and time. We could describe how the parts of each actor move relative to his other parts and his impact on his physical environment, including the stream of artefacts that are created, modified and destroyed. We could describe the emotional content of the situation and its complex interplay of feelings. But if we describe the situation in those terms, we deal only marginally with the fact that the actors in the situation think quite complex thoughts about it, that they are to some extent aware of the intentions of other actors over a wide range of conceivable situations, and that they interact with each other through institutions which have a fascinating combination of change and stability.

Whatever philosophical arguments we might have about the ultimate possibility of reducing our understanding of the complete range of human behaviour to some of these elementary levels, the fact remains that you and I are nowhere near being able to use the language of such reduced levels to describe all that is significant to us. Nor is human society going to be conducted in those terms in the foreseeable future.

So instead of using such terms to conjure up the totality of a situation and talking about actions, theories, proposals, change, and stability as though they were to be approached through other points of view taken to be more central, I propose that they should be put at the centre of our theory of reflective human action, and hence at the centre of our theory of doing OR. Together they constitute the notion of *action programme*. I intend the notion of action programme to be a point of view from which to regard human activity. Although almost any activity can be so regarded, I am *not* claiming that it is a point of view from which we can give a *comprehensive* account of the activity; I do not believe comprehensiveness is possible, so I do not suppose I am offering you a unified theory of life! Nevertheless, I have found it a point of view from which a variety of otherwise disparate problems appear to have common origins, and which therefore permits the ordering of one's experience over a wider range. Moreover, it is a point of view not too far removed from the fragmentary mix taken into one's thinking from the broadly scientific cultures in which OR people and other professional advisers are educated.

Action programmes are characterized by the actions that happen, the theories that the actors have about their actual and potential actions, and the proposals that are made within the programme and across its boundaries. The notion of action programme is not therefore approached through any intervening notion of the stream of consciousness of the actors, nor have we taken any view on the reliability with which the theories in the programme reflect externalities. Neither is the notion of action programme approached by way of any claim that the actors are, or ought to be, or ought to wish to be rational; we take the thinking and action of the actors as it is, whether it appears admirable or reprehensible, consistent or schizoid. On the other hand, by taking their thinking and action as it is, we do not intend to imply that it has internal unity or that in some mysterious way it authenticates itself.

The use of the word 'programme' is a little unusual, but a cursory glance at any dictionary will dispel the notion that language works by one meaning to one word. Here it is not being used to denote a set of intentions such as might be included in what a company calls its programme of future investments or in what a city calls its programme for a week's music festival; the word 'proposal' is the general word I am using to cover plans and intentions of that kind. Rather, the word 'programme' is used with the idea of a continuing process which has associations with the perceptions and intentions of human actors and which also has associations of stability and adjustability as it progresses. Why then not use a word like 'process' instead? A negative reason is that 'process' in scientific and industrial culture has strong associations of purely hydro-mechanical functioning; I have little use here for a notion that might be taken to imply that minds work in a simple stimulus-response fashion. I am sure that for many things they do! But I would regard you as perverse if you insisted on explaining yourself at your creative and imaginative best as simply responding to stimuli! It is a view which appears to be implicit in much that has been written about the systems approach, and it is a significant fault when that approach is commended for the design of arrangements for cooperation.

A positive reason for preferring the word 'programme' is that it is a direct extension of the way in which Lakatos used it to describe how research develops over time, sometimes over considerable periods of time. I need to anticipate some of the issues of later chapters to convey the force of the term as Lakatos used it.

Much of what I say later on the status of theories is a fairly straightforward application of the ideas of Popper, who among other things took the conjectural status of all knowledge as a central idea in his enquiry into the logic of increasing scientific understanding, and showed that such an apparently negative starting point leads to a more fruitful way of looking at the open-ended always-likely-to-be-challenged status of scientific theories than the older science-as-the-discovery-of-truth picture which it overturned. However, his early views implied that existing well-entrenched theories should logically be overthrown by a single counter example. This did not happen historically, and it was clearly right that it should not in many important instances.

What did happen, and what seemed to be logically well-founded, was that an adverse result would stimulate the reappraisal of all the explicit supporting theories on which the arguments rested, and might well cause newly-conceived articulation of implicit supporting theories that had not previously been recognized. Lakatos was led to formalize this view as the result of the apparent conflict between Popper's views and the views of Kuhn, whose concept of 'paradigm' provided an alternative and not too dissimilar view on the resistance to relinquishing ideas in current use in the scientific community. Lakatos used the conflict surrounding Newton's theory of gravitational attraction as an example of a *research programme*: the sequence of core and supporting theories through time together with the implicit rules for responding to adverse results and for selecting new applications. A first application of gravitational theory to the movement of the Earth round the Sun led to predictions which disagreed with accurate astronomical observations. A simple view of disproof would have led to the theory being declared disproven at that point. What happened over a period of years was that the theory was preserved as a core idea and prompted new applications and a series of amendments and supplements to arguments: for example, replacing an early model which had the Earth going round the Sun by the truer and slightly different model of the Earth and Sun going round their common gravitational centre, and replacing the implicit view that light travelled in straight lines by the corrected view that there was some bending from passing through the Earth's atmosphere.

Although the Popper–Lakatos programme is principally concerned with countering other programmes in the world of philosophy, it contains much that not only satisfies that purpose but also happens to be immediately accessible to a surprisingly varied group of people in fields as diverse as art and astronomy, who welcome the ideas as articulating and resolving some of the central problems of describing progress in their own fields, and who declare that their capacity for theorizing has been enhanced. The Popper–Lakatos programme goes some way to providing similar articulation and resolution for the world of OR. There is, however, a group of problems about purpose, choice, conflict, and ideas of what constitutes good and best, for which I found much that challenged me but nothing that finally satisfied me. It is for those problems, and for a number of other purposes, that I introduced 'proposal' as another key notion to stand alongside 'theory' as a pair of ideas which when used together provide discriminating power in some quite new directions. So my use of the word 'programme' includes not only the development of theories over time but also the development of proposals and the development of the repertoire of actual and imagined actions.

But in other ways my use of the word 'programme' represents a shift away from the spirit of Lakatos' work. For example, the theories he appeared to have in mind for the content of a research programme were a small compact set of key ones, as were the implicit rules. So my notion of including all the seething mass of theories, however well- or ill-founded, simply as they coexist looks like a riskily uncontrolled expansion, even though it is in the spirit of Popper's later idea of life being theory-saturated. And if moreover I declare that the actors are quite free to define for themselves how large or how small a slice of their life they wish to include in a particular programme, then it is clear that the observer is losing power to analyse as the participant is gaining it.

That may be so, but it does seem to be consistent with language as we use it. You have no difficulty in seeing your 24 hours-a-day life as an entity, nor of seeing the on-going sequence of interactions with the store where you buy your food as another entity, nor of seeing the various separate episodes that make up your reading of this book as another entity. So you are used to segmenting your experience at very different levels of aggregation, as seems necessary for whatever you are considering. By using the word 'programme' about whatever segment you choose, I am in the first instance reminding you that you can if you wish discover a great deal of meaning that you have ascribed or would ascribe if prompted in various directions. By 'programme' I also want to suggest that two important elements of your thinking are:

(i) Your picture of how things are and would behave under various future conditions,

(ii) Your picture of what you and others wish to see happen, or would wish to see happen under a variety of possible circumstances that might occur.

For example, among the theories that were active when I prepared this page were:

It will be published.
It will find its way to readers.

The readers will have a good vocabulary and/or a dictionary, and will be able to face abstract ideas.
It will be read by some readers before being used, along with the other pages, for other purposes.

Among the proposals that were active when I prepared this page were:

Series editor to me: Complete your book.
Family to me: Do not watch TV this evening.
Me to me: Do something to try to illustrate what you mean by theory-saturated and proposal-saturated.

Among the theories that were latent when I started to write this page, and were brought into consciousness only because I had to find some examples for you, were:

(If I heard slippered feet) *Coffee is on the way.*
(If the lights went out). *Something's tripped the earth-leakage switch again.*
It's a power failure.

Among the proposals that were latent when I started to write this page, and of which only the last has been activated, were:

(If I heard loud disagreement). *Me to children*: Make less noise.
(If the phone rings). *Me to me*: Answer it.
(If no coffee appears by 11 p.m.) *Me to anyone handy*: How about making some coffee.

In the same way that we can identify segments of meaning for our personal activity at all levels of aggregation, we can identify segments of meaning for our joint activity with others, again at all levels of aggregation. So, for example, we can agree that a game of cards is taking place and as participants we are aware of a common bundle of theories, like what cards there are in the set and what might be poor actions in particular circumstances, and we are also aware of a common bundle of proposals, like the rules of the game and the social conventions of how card playing proceeds in our part of the world. At the same time there is an extensive bundle of theories and proposals that might be brought into play by the turn of events: for example, if it is discovered that someone has the wrong number of cards, there is likely to be a variety of competing proposals together with a rush of freshly pictured consequences.

Not everyone has quite the same bundles of theories and proposals even in a simple situation; the differences may or may not be instrumental in causing the programme to develop in the way it does. For some purposes it may be useful to identify who holds what theories and proposals, but for other purposes we may not need to make that distinction. Again, for some purposes it may be useful to identify many of the theories and proposals that are not held in common, and for other purposes we may not need to go beyond what is common to all the participants.

In the same way we can identify segments of meaning for the joint activities in which we participate, we can learn something of the segmentations that other people have made for themselves individually or in groups. It might be that they pass their segmentation on to us quite formally: for example, 'We are pleased to announce the formation of a new subsidiary, Qzymt Ltd, to handle the production of zymase'. Our notions of the programme labelled Qzymt would then not simply be derived from the formal announcement but would almost certainly be supplemented by our existing theories about companies and an awareness of the extensive proposals under which they operate: for example, legislation on the disclosure of information and conventions on accepted commercial practices. We might, of course, be quite wrong in the picture of Qzymt with which we continue; it might be a programme within which a proposal to diversify leads after a few months to complete specialization in the breeding of zygoptera, or it might be a programme which exists only in documentary form for the next two years.

We are not necessarily aware of the programmes of others even though we may be affected by them. I suppose the classic example is the person who for several years participates in the foreign ministry programme of a country and concurrently participates in the intelligence gathering programme of some other country.

Another example might be the person who participates in a marriage programme and at the same time in an undeclared affair programme. Our understanding of programmes that affect us therefore ranges from complete ignorance to partial ignorance!

The existence of a joint action programme needs some degree of social agreement; indeed, following one of the core theories of Silverman's account of sociological understanding, we may say that the continued existence of a programme depends on the continued rehearsal of its meaning by those who participate in it or have transactions with it. Within that limitation, action programmes can begin, expand, wither, change direction, lie dormant, modulate by stages into something completely different, and end. Although they can, and do, exhibit considerable stability, there is nothing inherent in them to prevent them becoming completely other, and it would be a mistake to invest them with the properties of organisms—individual living entities whose development is constrained by their molecular biology.

On a large scale the various world civilizations can be seen as action programmes. Trade is an action programme. The local supermarket is an action programme. So is the banking system, a town, a government department, a protest movement, a learned society, a theatre company, the building of a power station, a flower show, a pop group, or a summer camp. For the participants of an action programme there is some sense of continuity through time. What gives it stability is that some of the theories or proposals or repertoire of actions are held to be central and only to be removed or relinquished under the greatest stress if at all. What gives it adaptability is that many of its theories and proposals and repertoire of actions are disposable or modifiable or replaceable by new ones: the closer they are to the core, the greater the pull to retain them and the greater the sense of partial discontinuity when they go.

The idea of an action programme bears some resemblance to the sociological idea of an institution and its charter, though I take the idea of action programme to differ in its inclusions and exclusions. But there is a similarity in the account of stability. Because the concept of action programme contains the concept of core, there is immediate access to the idea that its behaviour may be governed for long periods of time by the core and that it can exhibit purpose-like behaviour. Without the notion of core we are left with a model of a sequence of equally unresisted responses to the stimuli of externally presented theories, proposals, and actions: a model in which the actors are not asserting themselves.

The actors participating in an action programme cannot give you a full list of their theories or proposals or repertoire of actions on request. Many are of such long standing that they seem to be simple facts of life rather than anything to be articulated and scrutinized. Moreover, many of the theories and proposals that an outside observer might think were implicit in the programme might never have come near to being consciously perceived by the participants, and if something of what is implicit has been glimpsed it might have remained far from being fully articulated in language. However, almost any change in the action programme or its environment may lead to previously unnoticed or unexamined content being noticed and examined. There may then be cause for reviewing, modifying or rejecting some of the present content and introducing new content: but familiar programmes are extensively unarticulated, with much of their repertoire not reflected upon in any depth.

Reflection in action programmes is of necessity limited: a sudden increase in reflection before a particular action will slow down the action programme locally, though that might be regarded as worthwhile within the action programme if later stages are in prospect likely to be more satisfactory. If later stages are then in retrospect deemed to have been more satisfactory, it will give retrospective support to having reflected more. Retrospective assessment of reflection is however an uncertain matter, since little of the action programme of the time will have been recorded. In this respect, fast-moving self-defined patchily-articulated action programmes are an altogether messier object of study than research programmes, whose key sequences of developments look more like the clearly articulatable moves of a chess game.

When I wrote just now of a change in an action programme being regarded as satisfactory, I ought perhaps to have stressed that the satisfactoriness or otherwise would depend on the programme from which the change is regarded. A change which leads to increased military effectiveness in one group is presumably satisfactory to that group and unsatisfactory to its opponents. However, it would be wrong to see all programmes as being in conflict; some changes could be simultaneously satisfactory in many programmes. Yet at the other extreme it can be quite hard to find actions which can be regarded as satisfactoy in every programme they affect; even advances in medicine can be a loss to those who provided the preceding treatment, except perhaps for those among them whose own future treatment is improved.

The point of view I am suggesting is not quite so tidy as a view of systems related to subsystems and supersystems: there is less sense of hierarchy and less sense of necessary inner consistency, and there is more emphasis on discontinuity. Since actors participate in many action programmes concurrently there may be role inconsistencies which originate in the inconsistency of theories and proposals between different programmes, in addition to any inconsistencies within a programme. An actor does not necessarily notice the inconsistencies, nor feel bad about them if he does, nor feel any difficulty in choosing from the repertoire of actions. But such inconsistencies do include those which arouse feelings of conflict and indecision in the individual. Insofar as they are articulated, theories and proposals offer scope for reflection within a programme and for advisory intervention from outside the programme. The resolution or escalation of conflict between and within action programmes is itself effected through actions, and the announcement of theories and proposals, *not guaranteed to succeed*; so for example diplomacy sometimes leads to war, and factions within an erstwhile single group go their separate ways.

You may perhaps feel that to objectify action programmes in this way is to lose sight of the individual. For the present I will simply say that we live in a society where programmes are freely objectified as a matter of course in the normal use of language, and that many programmes are given legal status: the limited liability company is a conceptual invention of the nineteenth century which is regarded as quite distinct from any of the actors associated with it, even if to all intents and purposes the company is entirely controlled by them. Indeed, there is in English legal history a celebrated case of a company run by one man going bankrupt: he had, however, defined sufficient different roles in relation to the company that he also had in law a claim on the remaining assets before any other creditor and his claim to those assets was upheld even though his actions were the prime cause of the losses the other creditors sustained. Your response to that example might range from moral outrage to admiration of anyone clever enough to make legal concepts work for him so well. But the lesson is clear: programmes with legal status are an intrinsic part of our industrial society and will be operated as objectively distinct entities from the actors associated with them. The clever manipulation of whole families of programmes is now a standard part of the acquisition and retention of individual power in industrial societies, and the subject of continual legislation to try to exclude from the repertoire of permitted actions those that lead to reports of dissatisfaction in such other programmes as can command enough attention from the legislative programme, be it by parliamentary debate, lobbying, organized protest, or violence. There is also growing recognition that many governmental and quasi-governmental bodies are inadequately specified as programmes by legislation and are inclined to adopt proposals for their operations which lead to actions that do not have widespread assent.

The most widespread fault in the definition of programmes is that they are credited with properties that they do not and cannot have, usually because the limitations of theory-making processes and proposal-making processes are given insufficient attention. The result is proposals which are ill adapted to the changing content of progammes. Clarity on this point seems to me to be especially necessary for people whose chosen role is to intervene to give advice based on analysis. It seems to be not uncommon for members of OR communities to be asked to intervene to provide advice or artefacts with impossible properties, and for them to be induced, or even induce themselves, to see themselves as capable of doing so. There is in the end something dull and stifling about unjustifiable confidence in the rightness of what one is doing; I hope that in succeeding chapters you will see that an apparently negative view of human capabilities is a necessary basis for innovation.

3. Actions

Our own actions may be thought of, and examined, at almost any level of aggregation, from the slightest movement you make as you sit reading to composite notions like 'I lived in London for 14 years'. If we aggregate too much we may miss the significance of particular variations in what we have aggregated, and if we divide too much we may miss the significance of some overall pattern. The purposes we have in hand affect the degree of aggregation which we undertake and the degree of aggregation that would best suit us. We are aware, if we think about it, that we customarily aggregate actions in our thinking. There are many things we do not normally think of as composites of smaller actions unless we are stimulated to do so: for example, you do not normally think of going to the kitchen as a series of individual steps unless walking is painfully difficult for you or unless you are remembering an entry of high drama.

We are quite used to many skills passing from an awkwardly followed combination of several moves to something no longer reflected on. For example, I no longer have to think what to do with my fingers when I play the piano since most notes find themselves, but when I play an organ I am still conscious of what I play with my feet as a set of discrete actions rather than the single action of seeing the printed music and hearing the intended sound being produced. But even what I do with my fingers repays conscious analysis from time to time. So there is nothing immutable about the way we regard our actions, even though at any one point in time they might fall roughly into the three categories of *unnoticed*, like motor skills, *unitary*, like going to work, or *composite*, like the list of things I have to do by next week. It is quite possible by changing the way we regard a particular action that we can make changes in it that we value.

In a similar way, in action programmes in general there are actions which are so habitual that they are no longer noticed or considered in the normal course of events, other actions which are thought of as such but not analysed into any finer detail, and actions which are normally regarded as composites of many actions each to be undertaken with separate attention. In the past twenty years the analysis of the detailed actions of commercial transactions has been a painful experience in companies introducing computers to their commercial operations. Whilst there have been some spectacular innovations in some real-time operations requiring fast response, it has been common to have reports that the range and variety of a company's repertoire of actions has first been underestimated, then been partly ignored in the implemented system, and finally not been treated as subject to innovation.

How a particular action is regarded—unnoticed, unitary, or composite—depends on the action programme from which it is regarded. For example, the action of a month's bargain price sale in a minor department of a minor store in a company with many such stores would most likely be regarded as composite of many actions within the action programme of the department, a unitary action within the action programme of the store, an unnoticed detail within the action programme of the company, and a unitary action from within the action programme of a rival store. For customers' action programmes in their many thousands the action may be unnoticed, unitary, or composite according to the degree to which specific proposals for purchase exist in or are offered to and noticed by those thousands of action programmes.

Insofar as an action happens within an action programme which is part of one or several hierarchical programmes, it happens simultaneously in all the programmes but is differently regarded in each. Moreover, the action might have its own interpretation in programmes to which it relates only indirectly. We might well play a game of producing the indefinitely long list of action programmes that a single action might affect. Many effects might be small. But there are also well known stories of the for-want-of-a-nail kind which suggest that some actions can be seen in retrospect to have had a tremendously wide range of consequences, though the actions were not necessarily noticed as pivotal at the time.

There are a whole range of consequences for any kind of would-be-scientific advice giving. The range is quite open-ended: in the rest of the chapter I comment briefly on six issues which seem to be illuminated by the concept of action and action programme we have so far established.

3.1 Differential regard

Since the significance of a single action depends on the programme from which it is viewed, we have no reason to suppose it will be similarly regarded in each. So whatever we think about the ultimate possibility of harmonizing human affairs either relative to each other or relative to some absolute, we cannot suppose that it is practical at present to regard the various significances of an action as evidently reducible to a single measure which is translatable between programmes. Even if within a single action programme an action could have a single measure ascribed to it, we would need a vector of measures to represent the measures ascribed by each separate action programme—that is, a list with a separate measure from each of the programmes. In our Western culture we are used to assenting to the view that the measures from many action programmes can be reduced from a vector to a single measure because we have a number of very general and longstanding action programmes where it is customary for many of the actors to forget the essentially vector nature of measures. Indeed, when anyone talks of the best action to choose as though 'best' were an absolute property independent of the programme from which it is declared, then he has already made a profound mistake. We might stress this even further by declaring a formal theory along the lines:

> No pair of actions can be specified in which one is preferable to the other in all conceivable programmes.

Another way of saying it would be:

> The significance of any action is conditional on the programme from which it is viewed.

There seems to me to be an endless range of situations in which the idea of differential regard could be useful; social life is crowded with episodes where participants talk about significance as though it were an attribute of what is referred to rather than an attribute of their own regard. But I will comment here on just two general programmes: trade and law.

We have an extensive exchange-of-obligation programme which at one extreme can be entered by anyone in possession of some quite general symbols of obligation, like coins and banknotes, and at the other extreme requires formal and deliberate assent from within two or more action programmes in the form of a specifically negotiated contract. There is much that could be said about the exchange-of-obligation programme: for example, about the mass of widely agreed theories and proposals which give sufficient agreement for it to be a highly useful programme for long periods, about abrupt discontinuities caused by currency collapse or revolution or abrogation of agreements, about the difficulties that state treasuries have in keeping track of the scale of exchanged obligations and hence in influencing approaches to unwise commitment, about the difficulties that state treasuries would have of understanding the implications of obligation agreements even if they knew about them. But there are two more basic points which I want to make here:

(i) Exchange of obligation would be unlikely to occur unless the obligation was differentially regarded.

(ii) Because perceptions of the theories and proposals of the general exchange-of-obligation programme change relatively slowly, the money symbols in use at the general end of the programme have become widely accepted as appropriate measures for reducing vectors of measures to a single figure. Such measures are not remotely appropriate for some action programmes, but there is a tendency to use them widely and to propose their use even more widely, perhaps even to elevate them to the level of esteem once held by words like 'right' or 'good'.

We also have an extensive action-restraining and action-requiring programme represented by the extensive set of proposals we call laws, together with a well established network for monitoring and punishment, and together with the provision of arbitration between action programmes in conflict. The legal action programme rolls on more or less according to expectation, with a minority of action programmes rejecting some or other of the proposals in the general law programme (e.g. groups using violence in support of changes in the constitution of the state) and thereby causing others of its proposals to be activated (e.g. arrest, trial, and punishment). From time to time the legal programme is modified by those bodies which promulgate laws, by new interpretations offered in the course of making judgements, or by threats and inducement. There are periods when a legal programme is sharply altered to suit some action programmes with marked disadvantage to others: this may for example be the result of a violent change of government, or the result of dictatorial consolidation of power. If there were occasions when an action or a proposal was regarded as lawful in all action programmes we would have consensus. More usually an action is differentially regarded as to its lawfulness or its equity in different action programmes. At its mildest, differential regard then leads to proposals such as placing some restraint on when and where certain actions can take place (e.g. restrictions on selling alcohol to children) and at its extreme to social breakdown (e.g. civil war).

3.2 Choosing the depth of analysis

Some of the most successful OR interventions in action programmes have been of the kind where a new view has been taken of the extent to which actions should be paid attention to, regarded as unitary, or regarded as composite. For example, computer methods might usefully allow a distribution planner to publish detailed route recommendations in addition to the list of destinations, or the control of development in a mine might be improved by defining smaller tasks. Conversely, a scheme of reporting by exception might screen out just those parts of a programme that need attention instead of the previously customary report on everything.

In intervening in an action programme, particularly a high-level one, the perception of actions in that programme is often checked-out by quite extensive examination of the perception of those actions in action programmes lower in the hierarchy: for example, what appears to be a simple unitary action of letting a contract at a high level might appear to have a number of unexpected properties when viewed by the people who will be in the thick of implementing it—indeed, it is quite common in my experience for senior people in an organization to be significantly unaware of some of the consequences that stem from the instructions they issue. The consequences of examining how an action looks from different and preferably contrasting, programmes can among other things be that attention to action is introduced, reclassified, or withdrawn, and that attention to actions takes place according to some new format of proposals for the frequency and method of paying attention to action. By paying attention to action I do not simply have in mind the formal information systems of management but also what is done in response to the information.

Another simple gain from looking at the perceptions of an action in many programmes is that the advisory programme thereby has a potentially rich source of competing theories which can be critically compared and which may enter the sponsoring programme.

Evidently an intervention programme has an indefinitely large range of possibilities about the degree of depth in which to examine past, present, and imagined actions. Some OR programmes have had a considerable history of the deep examination of actions from the point of view of action programmes lower than and external to the sponsoring programme: some of these OR programmes clearly have a core proposal that deep examination will take place—this is characteristic of several in-company and in-industry OR groups in the UK such as those in coal, steel, rail, air, and defence. Other OR programmes have been oriented quite deliberately to taking the theories, proposals, and repertoire of actions more or less as they currently exist in the sponsoring programme and to offering methods of examining alternative proposals within those given perceptions—this has characterized bureaux services in the application of mathematical programming and has also characterized some approaches to high-level planning in government where providing a well-articulated model with extensive internal coherence has been seen as needing attention before refinement of inputs to it.

In general there is no ground for stating that it is in the interests of an entered programme for the examination of actions to proceed in a particular way. There are potential, unknown but perhaps guessable gains and losses however it is done: the actual gains and losses depend both on the characteristics of the entered programme and on the characteristics of the intervening programme. So if we are invited to consider advising within a programme, it will be inadequate simply to say that we will carry out an OR study, since that will not give any indication of the main line of attack and may confirm both ourselves and our sponsors in the mistaken view that there is for any one problem a knowable correct line of attack. There is no such thing as a right way of tackling a problem, though I would in any particular case regard some ways as more promising than others. In effect what I do in a particular case is to try to imagine what changes in the entered programme might have taken place by the end of the intervention: will its repertoire of actions have been altered, will its bundle of theories and proposals be in some way more functional? I also need to answer quite specific questions on what actions I propose to take towards making a contribution by my intervention. Yet, because it is a venture into the unknown, the intervention cannot be firmly planned: I have to be prepared for surprises and for changes of direction and changes of imagined end-point.

3.3 Extending the repertoire of actions

A sponsoring programme might well expect an intervening programme to suggest new forms of action. There has been such considerable attention given to the problems of devising abstract technologies for choosing between a given set of actions that the OR communities have at times seemed to be retreating from the clear and frequent requirement of sponsoring programmes to conceive novel actions.

I say *conceive* them rather than *deduce* them because novel actions cannot logically follow from any theory deductively; that is, novel actions cannot be found by logic, mathematics, or computation. They must first be conceived. After that we can, if we wish and are able, develop a theory about the now-conceived actions. Moreover, if the actions have been conceived in the form of a general set of similar structure we might well then use mathematical methods once or repeatedly to select from the now-conceived set. On the other hand, it may be that no general theory can be forthcoming without specific data collection. For example, once you have had the idea that hens might lay better with background music, you can carry out a programme of measurements of egg-laying as it is affected by the loudness of a piece and how often it is played; then you can even construct a little economic model for yourself to find a point of balance between increased costs of music and increased proceeds from eggs!

Although that is a light-hearted example, there are quite serious examples of limited conception of actions. For example, a great deal of advice on the control of inventories has been given on the basis of models in which it is supposed that the only action open to a store programme as inventory falls is to place a normal request for more and wait: there is in the models no picture of action to urge things along, no picture of preferential delivery, no picture of the possibility of discriminating between different classes of demand for what is in the inventory.

The conceiving of novel actions seems to be relatively rare: I think it unlikely that more than a handful of any OR community can rely on a continuingly fertile imagination simply as a result of confronting each problem. Much of what is new to a programme is likely to be introduced or induced by interaction with other programmes. The model I have here is not one in which the intervening group *knows* which programmes to interact with but one in which the intervening group consciously elaborates its arrangements for interaction. A wide range of specific recommendations fall into the category of deliberately created interactions: examples include

(1) The formation of OR groups with varied backgrounds, the formation of intervention task groups jointly between entered and intervening programmes, locating members of the intervening group in the goings-on of the entered programme,
(2) Moving freely among all the programmes concerned with the intervention and at all levels, engineering greater freedom of access for people than is usual,
(3) The regular practice of presenting the intellectual content of the intervention to others whose habit of asking for structured understanding and of imagining operational weakness may induce new responses—I say induce because the experience of having several teams reporting to me regularly on their interventions was that they left with new ideas which had emerged from the rapid interplay of suggested understanding and criticism with no one quite sure about the exact genesis of the ideas,
(4) The freer use of literature search, but perhaps more importantly the use of contact networks to locate those with relevant experience or understanding of somewhat similar problems—this is now much more important than in the early days of OR since the preponderance of studies are not classical.

But none of these examples is intended to give the slightest support to the mystical notion that if you have a problem there is someone somewhere who knows what you should do. However confident the advice, it is at most conjectural that you will in retrospect be satisfied you took it!

3.4 Limits to prediction

As a consequence of the possibility of novel actions being conceived, the course of an action programme through time, and of any measure of it, is not predictable; nor is it possible to say reliably in advance, in probabilistic or any other terms, how near to being correct any predictions might be.

The possibility of novel actions is a sufficient ground, although by no means the only one, for rejecting the claims of anyone who believes himself privy to sure knowledge of how human history will turn out! And whatever view you take of the ultimate predictability of the universe in principle, it would be foolish to suppose that this offers much hope of practical advantage now except insofar as it gains you whatever advantages there are to be had from being admired for your certainty: these can be considerable!

In addition to the scope for conceiving and introducing novel actions within a programme, it will also be acted on by a host of other programmes, some of them not yet in being. Some of the potential actions are pivotal: for example, massive nuclear exchanges, chemical–biological warfare, or the release of artificially constructed viruses by mistake, and many kinds of less dramatic but no less effective political, military, or economic action. And this is quite apart from any unexpected natural phenomena which may affect programmes. So, short of an incredible increase in our capacity to detect, record and analyse programmes, our ability to predict them can draw no support from the kind of presumed regularity in things which we now feel we can rely on in the descriptions of the physical world given by the natural sciences. Rather, our ability to predict action programmes depends on their kindliness or wildness in changing steadily or suddenly.

From this point of view, action programmes are so far removed from the regular basic mechanisms of nature that a kind of quasi-stability is the best we can hope for. Evidently, if an action programme contains a core proposal that some particular measure of it will be held stable, it is quite possible that the measure will be stable for long periods until that is no longer possible and the proposal drops out of the core or the programme terminates: for example, a bank may pay a steady annual dividend each year until it crashes. One can also note that if a very general programme remains reasonably stable in its theories and proposals and its repertoire of actions, then it may well be possible to develop quite good predictive theories about lower-level programmes, subject always to the risk that the predictive value of the theories might suddenly deteriorate if the higher-level programme changes. So, for example, the exchange-of-obligation programme changes sufficiently slowly for predictions about economies and industries to have at least some local validity; but the invention of what became known as the eurodollar led to changes in control and ownership on a quite unanticipated scale, and a fourfold increase in oil prices was widely not anticipated.

Adaptability of action programmes to the unforeseen is therefore a constant requirement, since changes in its environment of other programmes continually move it into a situation not quite like anything previously experienced, and possibly with significant discontinuities. Insofar as the new situation can be satisfactorily identified by the current bunch of measures in use, and insofar as the new situation causes and implicitly requires no change in the bunch of theories in use and in the bunch of proposals in use, and insofar as the new situation generates or implicitly requires no change in the current repertoire of actions, adaptation can be achieved by an abstract technology.

However, an action programme is not limited to adaptation. It can be a source of innovation in its own right, and it can include contingency plans for whatever futures can be imagined.

3.5 The inadequacy of conventional quantitative modelling

Since action programmes depend crucially on theories and proposals, the approach that models measures of actions and measures of consequences through statements relating how the measures push and pull each other, so powerful in the natural sciences, is by its very structure of formulation a shallow approach to the modelling of action programmes. Such theories are an extremely useful tool of engineering and adaptation within periods of quasi-stability. But they have little or no claim to be in-depth theories of programmes. I take OR and all other quantitatively oriented approaches to intervention to share this shallow nature at present, and to be basically tools of trial and error which will serve us well when things do not get rocked too much but will leave us perpetually behind when things are rocked. The in-depth understanding I have in mind would necessarily include an account of the dynamics of theories and proposals as well as actions and their consequences.

If you think this is too sweeping, think of the extraordinarily turbulent history of human affairs, with civilization rising and falling over as little as a few hundred years, with organizations and states changing and growing and ending, with means of transport regularly succeeding each other, and with immense changes from military activity. Even in apparently quiescent periods, programmes are continually impacting on each other and being modified. It seems unlikely that we shall attain any deep insight into the relation between actions and their measurable social consequences if we ignore the intermediate role of theories and proposals, quite apart from the great potential for changing the relationship between actions and consequences by directly affecting the theories and proposals in other programmes by, for example, advocacy, advertising, propaganda, or by control of the language which can be used in the mass media.

I make this point not to be depressing but to counter what seems to be an overestimation of what can be achieved by studying the relationships between those features of society that currently have quantitative measures attached to them. It is sometimes almost as though there was a belief that, because the validity of logic seems to be permanent, the validity of premises and conclusions must also be permanent. However, the validity of premises about the real world is always open to refutation particularly if, as I have argued, there is continual change of some kind, which is in turn likely to have the world departing from once valid premises. However, I see no need to strain over metaphysical points about whether science is a way of approximating to truth or whether it is purely instrumental. Although there is no way of proving it, I have little doubt that natural science is close to some kinds of general truth as well as being instrumental. And I have no doubt that OR is 'making do' with models that are largely instrumental and only temporarily close to the truth. With so many societies crammed into so short a history it seems folly to think otherwise.

3.6 The necessary impermanence of recommendations

We must regard any discoveries about programmes as being only temporarily supportable if the programme is to go on developing through time. That leaves open the question of how temporarily, but it implies that any model that we propose for ongoing use, or any realization of a particular abstract technology we propose for ongoing use, is in need of checking and revision in a permanent programme for keeping instrumental theories and proposals and action repertoires under review.

At a more general level, and recognizing the inherent limitations, we may be able to observe and classify the changing species of action programmes, to try to learn how we can better describe them and so make our engineering responses more reliable.

I expect and hope that we shall also continue in the classical tradition of intervening in programmes of which there is only one example which modifies and adapts and innovates through time. In doing so, our aim is to clarify and criticize the theories and proposals and repertoire in current use, to extend the repertoire, to provide theories of greater reliability, and to imagine proposals of greater effectiveness. Surprisingly, that seems to be what members of the OR community manage sufficiently often for it to be a tenable view. It can be frustrating and exposed, and there is no guarantee of success beforehand, but it can also be very satisfying.

4. Intervention

Since intervening individuals are so varied and since the programmes entered are so varied, we can hardly regard OR as a commodity of uniform specification which can be bought and sold. Although we can mean something sensible by 'OR in Finance', we need to remind ourselves from time to time of the human situation: that in a specific instance Jim Smith and Penelope Roquefort of the Operational Research Agency of Cardiff, London, and Edinburgh will be working closely with John St John Fortescue and Bob Brown of Llombos Bank. The pair from Oracle and the pair from Llombos will also continue as participants of the many programmes which are already discernible in their parent organization and in their private lives.

If the intervening programme is limited in its description of the entered programme to those actions, proposals, and theories already explicit in the entered programme, then many low-level actions will remain unexamined, many theories will appear to relate in odd ways to actions and their consequences, and many latent proposals will remain undiscovered since they may not be indicated by actions if they are prohibitory or if they are conditional on rare events. Nevertheless, it is with its present content that the participants of the entered programme will receive and assess the intervention, though the interveners and the sponsors expect the intervention to change the content in some significant way. It would be self-deceiving if the participants in the intervening programme expected their contribution to be judged in their own terms: for example, if the intervening programme sees itself as establishing truths by scientific investigation, then even if the entered organization is unhealthily likely to accept whatever would-be-truths that are handed out, there may still be nothing of operational significance for it.

The contrast in content between the entered programme and the intervening programme is a potential source of grievance. It is common for the two programmes to be substantially more different than the participants of either programme realize, and for the broad outlines of the likely course of an intervention to be seriously mispredicted. A part of the problem is not simply that the theories and proposals and perceptions of action repertoire are different but that the very words and images that go to make up formal statements are different: both programmes are to some extent without a sufficient language for considering the other. Yet for the proposed intervention to have been seen as worth encouraging, there must have been a view in the sponsoring programme that the two programmes were sufficiently similar and sufficiently contrasting for there to be interesting potential from an intervention. So it would seem that in the preliminary mutual appraisal significant differences of content can be overlooked. For example, the entered programme may have extensive proposals which limit the disclosure of information, even at quite high levels in the managerial hierarchy, and this may be at odds with an OR group's unspoken supposition that the first outputs from a monthly planning model will be able to be subject to a wide range of criticism; I am aware of one case in which a perception gap of that kind led to a year's work being scrapped.

The intervening programme can be regarded as staying separate at its interface with the entered programme or alternatively having with the entered programme jointly created a new intervention programme. Each view emphasizes one or other aspect of separateness and togetherness during the intervention. It is perhaps conventional to think in contractual terms of just two programmes, but for the purpose of understanding what is happening to the bundles of perceived theories and proposals it seems more useful to regard the intervention experience as a separate programme. One is then likely to pay attention to differences between the three programmes. Occasionally that can be crucial. For example, the intervention programme can develop a much wider range of content than the participants realize, and thus create surprises at the point where it had been thought that all that was needed for the rest of the entered programme was a brief report: it may instead need a considerable appreciation and training programme in order to translate advances in understanding and repertoire from the intervention to the rest of the entered programme.

With that brief outline of some structural points about intervention, which was partly a reminder of the autonomy of programmes, I turn in the next four chapters to increasingly detailed examination of the content of programmes and what we can say about its validity.

5. Reflection before action

The point of view so far set out emphasizes the cognitive and imaginative content which can be articulated and which is subject to change, but does not deal directly with the argumentative activities or other mental functions by which the changes take place. There is much about the activities of reflection that we do not understand; there is therefore much that in prospect could become part of the set of theories that we actively use. But in trying to formalize what is implied by would-be-scientific advice, the contrast with everyday reflection and behaviour is heightened. In this chapter I comment on that gap and on other issues associated with articulating reflection.

5.1 A choice of viewpoint

There is a whole range of reflection before action. At the extremes are two idealizations, neither of them attainable: at one extreme there is reflection which has no hint of theoretical content however ill-founded; and at the other extreme there is reflection whose entire content has been exhaustively tested to the limits of human critical ingenuity. Perhaps more simply the unattainable extremes of reflection can be regarded as the completely mindless and the known-to-be-almost-certainly-correct. In between lies our actual reflection on what Oakeshott terms the imperfectly imagined futures we wish to happen or we wish to avoid: futures about which what he calls our deliberations with ourselves and our persuasive arguments with others take place in far less completely articulated ways than the published argumentative output from research programmes in mathematics or the natural sciences.

My notion of action programme is a framework suggested by the traditional aspiration of scientists to produce deliberative argument that is well-articulated and well-founded, and by implication either to produce persuasive argument that is similarly well-articulated and well-founded or to do away altogether with the need for persuasive argument. It is therefore at odds with much of life as it is lived. There is a not unsimilar tradition in the humanities that ideas should be well-articulated and well-founded, though in areas like law no-one would doubt the need for persuasive argument. The two traditions mix uneasily when the subject matter is reflective action in society: it seems that there are members of both traditions who suppose that there is, or that there will eventually be, a part of their own tradition before which the other tradition must cede its claim to primacy. I would rather view them as variants of the single tradition of articulation. The difference between the variants lies not so much in the aspirations as in the means of trying to achieve them and as in some of the subject matter addressed. They together form a tradition with a degree of articulation which distinguishes them from the general goings-on of society. It is the general goings-on of society into which the OR communities intervene to provide more completely articulated reflection; that is, the OR communities offer articulate intervention.

There are three further considerations which make it seem reasonable to adopt a point of view which emphasizes formal articulation. Firstly, the conventional abstract models of OR are *in intent* fully articulated announcements of theoretical content on the basis of which action can be chosen: the fact that they are almost all structurally incomplete in being without a model of themselves does not vitiate the intent that they should be free of ambiguity or, at a higher level of abstraction, the intent that such ambiguity as remains shall have unambiguous statements made about it. Secondly, it is a conventional aspiration that the output of understandings from an OR programme shall be transmissible and that they shall be able to withstand articulate well-structured critical debate; perhaps at this point, however, I am guilty of trying to put an unwarrantably respectable gloss on what seems to be the widely held belief that in some mysterious way the output of a would-be-scientific programme will be authoritative beyond the reach of any responsible criticism! Thirdly, whenever the actors in a would-be-scientific programme use language about the unarticulated content of programmes they raise that content to the level of language, thereby articulating what was unarticulated; in making that point I do not intend to imply that their only impact on the entered programme is through articulation, since they will also have an impact through many actions that are the subject of little or no thought.

That may account for my emphasis on articulation, but it does not account for my considering only three very broad categories from all that is articulated: theories, proposals, and imagined actions. Actions I have discussed already. Theories are intended to be at all levels of generality from the general theories of natural science, through the diagnostic statements suitable to a particular intervention, to statements of such simplicity that they would normally be thought of as facts rather than theories (e.g. the paper on this page is white). Similarly proposals are intended to be at all levels of generality from ethical proposals, through plans on what to do in some specific circumstances, to proposals of such simplicity that they do not need saying (e.g. when you get to the bottom of this page transfer your attention to the top left-hand corner of the page with the next highest number).

The three ideas are not independent of each other, and it is a fairly easy matter to make any one of the three seem more important than the other two:

(1) Actions are the main category, since each theory and proposal that I announce is an action of utterance; utterance is but one of my repertoire of actions,
(2) Theories are the main category, since I necessarily have pictures of the consequences of my acting and making proposals; I also have theories about much else,
(3) Proposals are the main category, since every statement of theory is an injunction to me and to others to use language in that way about that topic, and every action taken after reflection was first a proposal formed under the influence of countless other proposals.

So why choose those three ideas? Well, as in all uses of language, I want to make some distinctions and contrasts, and I do not think that there is in principle a tidy compact set of the only generally useful and true things that can be said about reflection. But I think that in the terms I am suggesting I can say several things that you will recurringly find useful. Moreover, I am trying to say them in ways in which they will stand up well to any critical appraisal of their validity, in ways which validly refer to my very act of writing them, and in ways which make it possible to maintain a distinction between validity on the one hand and preferences about subject matter and choice of viewpoint on the other.

Perhaps it would help if I summarized some of the considerations that seemed important to me at various stages:

Why actions?

A set of alternative actions is one of the primitive notions of OR literature, standing at the end of diagnosing a real system and at the beginning of using an abstract technology.

Actions are observable by others without requiring verbal self-disclosure by the actors: in that sense they are what is to be understood.

Why theories?

Theories are the dominant notions of the scientific world, in which there is a hope that eventually no other category will be necessary.

Theories in the scientific world and the everyday world do not differ in kind but only in the degree to which they have been critically appraised; the continuity between the two worlds provides part of a basis for understanding would-be-scientific intervention.

The understanding of action is poor without the actor's interpretation of it or without what we can imagine his interpretation to be.

Why proposals?

Social experience can be seen as a network of transactions in which suggested actions are accepted, ignored, or rejected.

Choice of action is profoundly affected by long-standing proposals, some of the most influential of which retain their power by being widely regarded as theories; the confusing of the two is one of the major sources of muddle in our time.

Theories are passive; no action follows from them.

Why no more categories?

Sometimes insight is gained by distinguishing between things previously regarded as similar; sometimes it is gained by seeing similarity between things previously regarded as dissimilar—using the word 'proposal' to cover so many different things is an experiment of the second kind.

5.2 Components of articulate reflection before action

During the course of articulate-reflection-before-action one can be aware of drawing on, or creating or modifying, members of three largely undeclared sets:

A set of concepts of possible actions,
A set of theories about what consequences follow particular actions, and perhaps even about how the actions lead to the consequences.
A set of proposals about which actions and theories and consequences are to be considered and in what light.

There will be a sense of incompleteness if any of these sets is unrepresented or poorly represented in what has been offered as articulate reflection to precede action.

That might seem evident. Yet it is surprising how often one comes across studies which seem to have been limited to reflection rather than reflection-before-action. That is, it appears that the real system has been studied only to make would-be-true statements about it, and there is neither much of an appraisal of the actions open to the actors in the real system nor much sign of do-able proposals for change or continuation, except perhaps for the hopeful assertion that the study shows the need for much more study! With an orientation to reflection rather than reflection-before-action it is left to others to make the connections.

It is an orientation which is, however, deeply rooted in the intellectual traditions of the UK. The aim of purely scientific activity is the development of general theories which correspond well with those classes of subject matter which are covered by the theory; this aim is well satisfied if statements of sharper discriminating power emerge to describe what is. It is a poor model for scientific intervention:

(1) The study of a single system will only provide a sharper view of what it is currently like, will not of itself provide any view of what the system could be like nor of how to get there, and will lead to interesting ideas of what the system might be like only if it prompts someone with a lively imagination.
(2) The conventional mode of reporting a study is to quote conclusions rather than proposals.

I have to press graduate students whose background is mainly an education in science very hard indeed to go beyond the scientific tradition and to think themselves into the role of being creative social technologists. 'Where are your proposals?' can sometimes make a study seem suddenly incomplete. But the person who is strongly rooted in the scientific tradition faces a possible problem the other way; if he produces imaginative suggestions he can be uncomfortably aware that they have not been drawn out of his enquiry by the style of argument that supports would-be-true conclusions about the system as it is, and without that sense of support he can feel his position is bogus. Perhaps then the present educational programmes in science need changes to make them suitable as a preparation not only for science about general classes of subject matter but also as a preparation for science for action in particular situations. The typical mode of examination is clear: a diagnostic report on an intervention into a particular situation with recommendations for action!

Articulate interventions are almost always into programmes in which the components of articulate reflection before action are of mixed status. Actions considered might be freshly conceived (e.g. introducing a new form of insurance contract), or might be from a widely available and perhaps unimaginative repertoire (e.g. adding a line of goods already widely sold by other stores). The theories might be little more than folklore (e.g. unemployed people are work-shy), might be dreamed up on the spot and accepted as plausible (e.g. if we move out the milling machine there will be room for two automatic lathes), might be new but have survived vigorous criticism (e.g. we got the lawyers on to it as soon as it happened and they are certain we aren't liable), might be one of the longstanding theories of a professional group (e.g. that an adequate safety ratio for some kinds of civil engineering structure is 10 : 1 compared with what would be needed if the materials behaved like idealized lab structures), might be the fashionable wisdom of the day (e.g. property is the best thing for investors to put their money into), or might be well-supported theories of science (e.g. theories of heat flow). The perception of consequences might emerge from complex argument (e.g. working out the consequences of changes in taxation legislation or in accounting methods), might be taken to be almost trivially indicated by what are taken to be similar situations (e.g. that reducing planned deliveries from five a week to three a week will reduce transport costs), or might be wishfully at odds with the evidence (e.g. to suppose that a plant agreement means the end of stoppages caused by disputes). Finally, proposals about what is to be considered and how it is to be regarded might be taken without question from prevailing convention (e.g. that of two otherwise similar investments the one showing greater profit should be chosen), might be a matter of personal conflict (e.g. a boardroom row over company policy), might be a matter of social conflict (e.g. in whose terms is the use of Indian territory for a pipeline in North-West Canada to be judged, there being no place in Indian thinking for ownership rights over land to be assumed and then sold), or might imply an impossibly wide open-endedness (e.g. that some action should only be taken with the full agreement of all concerned—such a proposal is less troubling than might logically be expected, since the person who utters it is then likely to show that he has a conveniently limited view of how to recognize full agreement and of how to discern those who are concerned!).

The role of articulate interveners then seems to be:

(1) To join the entered programme in articulating and elaborating the components of reflection before action,
(2) To reflect on the quality of the components, and to act to improve them,
(3) To participate in deliberative argument, and to elaborate the means for deliberative argument,

though these may be limited by the perceptions or the wishes of the participants.

5.3 Explicit, latent, and implicit content

Much of the conscious attention to impending action in everyday matters does not at any point have the content of its components articulated clearly and deliberately. Under questioning about a short episode of reflection one can distinguish images relating to all three components, but it is then not clear whether at the time of the reflection they were explicit, or whether they were latent in the sense that they were some of the available mental constructs that were accessible to the actor at that time, or whether they were implicit in the sense that they seem to the questioner to accord with the explicit and latent images but not to have been among the available mental constructs.

From the point of view of wanting to learn something useful about articulate reflection, boundary distinctions between the two are not particularly important: a teacher or a political activist would have much greater interest in trying to raise the various levels of consciousness they are interested in! They do, however, at the centre of their meaning have different implications for articulate intervention.

We might be impressed by the tremendous range of latent content we carry and produce so effortlessly. For example, a twinge of toothache can readily be associated with theories of tooth decay over time, of means of arresting decay, of the sequence of pleasure and pain under various alternatives, of professional competence to use the means, of who is likely to pay, of how to cause treatment to occur, and of the consequences of not being where one would otherwise be at the time of treatment. The same twinge can also readily be associated with proposals for the use of the telephone system, for the behaviour of patients, for professional conduct, for seeking legal remedy when things go wrong, and for acceptable domestic conduct until a remedy has been found.

Even for an individual the latent content of his reflection looks indefinitely large. By that I intend to convey something rather like what a mathematician means when he uses language about abstract systems: that however large a set of latent images we manage to articulate there will always be some more we have not yet articulated. The difference is that I offer it as a theory about real people rather than about an abstract system. The consequence is that in articulate intervention we necessarily deal with a subset of latent theories and proposals, and it is therefore always possible that some significant alteration will be needed to any model devised for use within the deliberative argument of the entered programme since, however simple the programme looks, the actors by whom it is held in being are capable of injecting almost anything into it. Since at most an articulate intervention and the realizations of abstract technology it leaves behind can deal with a finite structure of possibilities, the most ambitious hope for intervention is that it will provide all that is needed for deliberations on a temporary issue and that it will for a while provide all that is needed for deliberations on a permanent issue.

In contrast to latent content, the implicit content of reflection can look embarrassingly crude and evasive. For example, in discussing how to reduce the working capital needed, Arnold might say: 'If we get the invoices out a day earlier, that should lead to payments being made a bit sooner', to which Brian might retort: 'That's alright in theory but it won't work in practice'. In making a reply in those terms, Brian appears not to have noticed that it is structurally inconsistent, since what he is in fact contrasting is:

> Theory 1, held by Arnold: The idea will work.
> Theory 2, held by Brian: The idea will not work.

He also appears to have introduced a false contrast: that Arnold is theorizing and that Brian is not. I know that Brian's reply can be regarded as being no more than an idiomatic use of language which should not be taken literally, but is that wise if it offers the false and divisive concept that some people theorize and others do not? Are we to honour unconscious theorizing more than conscious theorizing? Perhaps surprisingly, that is an issue that merits some reflection; I will take up this topic in a later section.

In contrast to latent content, the implicit theories that seem to me to be significant are relatively few, and are of very general application. For example, we do not often rehearse the extent to which we assume there is a reliable continuity to reality: I do not expect the floor to give way when I start to type this next w . . . aaaaagggghhhhh! I jest, but with serious purpose, since we rely every moment on the continuity of most things around us, and at a much less intense level we rely on the continuity of the social milieu we live in but only realize how extensive that reliance is when change makes us notice. Because implicit theories underpin so much of what we do, rehearsing them can be found offensive, even though they are sometimes relied on erroneously, and even though their declaration might lead to the resolution of some problem on hand: for example, that we may not have seen what we think we have seen, or that we may not have understood as intended when we have heard an utterance that seemed clear to us.

In articulate intervention there may be understanding to be had in announcing some of the things that seem implicit to the intervener, but the understanding may not be well-received!

5.4 Emotions

Is not the action programme point of view rather too civilized? No and yes. No, in that actions include cruelties of the worst kind, theories include the most callous assertions of the worthlessness of others, and proposals include orders issued with the most vicious threats. Yes, in that the red-blooded world of feelings does not find a place in the ideas that make up the framework. That was part of the act of choosing a point of view.

In effect I started with an idealized world of passionless science, which I took to be the scientific world's idealization of itself. I then found passionless theorizing incomplete for acting in a social world, and declared that a minimum would be an idealized world of passionless theorizing and passionless proposing; and as an idealization of articulate reflection before action I supposed there was no need to go further.

But the formal description of the components of reflection has an air of neatness foreign to the actual processes of reflection in which a complex sequence of images passes through the consciousness of oneself and others, arousing and interacting with whole networks of feelings and non-conscious mental activities, and leading to some images being associated with particular feelings when they are recalled. I would like to have an articulated understanding of all that: it would then become a part of any programmes I participate in—some of its most important theoretical content. So my aim would be, for articulate reflection before action, to understand my own emotions and those of others. But as is true for everyone else, my actual behaviour will only partly be influenced by articulate reflection!

There was, however, another reason for omitting emotions from the basic framework, and that was that it seemed to me that a clean break was needed from previous attempts to include them: in the would-be-scientific approach to decision-making, for example, emotions had formally entered as preferences, but that seems a rather limited notion, and one which suggests somewhat passive membership of an established economic order rather than life where power is to be taken and kept by all kinds of forceful action. If the news media reported that most people were satisfied with the programmes of which they are members, I would think it reasonable to proceed from there by considering their preferences: since many major action programmes appear to proceed in ways which conflict with the feelings of individual members (e.g. large-scale political imprisonment, unwanted unemployment, compulsory education, high levels of taxation on high incomes, wars directed at civilian populations, liquidation of dissident groups, drab and damaging working conditions) I wanted a language in which the worst could be considered and in which there was free acknowledgement of the place of language in how things happen.

Emotions undoubtedly have to be taken into consideration. They are inseparably linked with mental processes, modifying our readiness to think in certain directions and blocking the possibility of some mental sequence altogether. Indeed, I take that remark sufficiently seriously to have intentionally made it possible for you to see some of the emotions that were linked with my writing of this chapter. Not too many, since that would violate to too great a degree established proposals for the content of intellectual writing, and that could in turn release feelings I would prefer not to stimulate!

There are countless situations where an understanding of physiologically or psychologically describable aspects of reflection and communication will give you greater ability to influence goings-on towards some desired consequence, or will prompt you to change your view of the sort of action to consider. For example, rather than just regarding negotiations as being between two given programmes, you might set them in the wider context of non-articulated ways of influencing programme content, right down to the fine detail of behaviour which normally stays well below the level of the conscious perception of the participants. At a more general level we shall regard the quasi-stability of society as not simply the result of negotiation between conflicting programmes but also as the result of the unconscious or deliberate influencing of the emotional attachment to particular ideas: texts on revolution and counter-revolution are centrally concerned with the manipulation of feelings other than those of the manipulators.

So there is no question of ignoring emotions. It is a question of from which point of view to regard them. The role of articulate intervenor implies that you articulate your understanding of them, but it is a role you will only be granted if your articulation generates favourable emotions in the participants of the entered programme.

5.5 Articulation: a controversial benefit

It seems that as a species we are content with poorly articulated reflection for much of what to do. There is no reason to suppose that there is a generally felt need for consciously increasing articulation in deliberative argument, nor that the highly articulated offerings of would-be-scientific interventions can take their place comfortably in the general goings-on of society. It seems that, in deliberative argument, challenges to particular utterances come thick and fast, and that it is not normal to conduct deliberative argument by the conscious building up of an elaborated internally consistent bundle of articulations and by the systematic and imaginative criticism of their content *against reality* as it is known.

In the circumstances, the extent to which would-be-scientific intervention is welcomed is surprising. Expectations of it are perhaps partly encouraged by analogy with the exceptional success of the natural sciences as an irreplaceable component in the provision of a stream of uniquely advantageous artefacts in the past two or three centuries. The hope is for similarly uniquely advantageous insights into the goings-on of an organization and its environment, or at least for uniquely advantageous tools for collecting and manipulating understandings of the goings-on and its environment.

There is, however, no *guarantee* that an increase in articulate reflection will improve an outcome compared with what it might have been without the increase. By the same token, there is nothing inherent in articulate intervention that guarantees a more desirable outcome. For example, in a study of the work of a particular materials controller I found it impossible to devise rules which would lead to better performance than was being achieved—the operation was small enough for the controller to know more about the materials and their place in the goings-on than it would have been feasible to capture by any permanent realization of an abstract technology that I could imagine, and the conventional abstract technology of the time certainly performed less well. And there are plenty of instances where articulate intervention has led to erroneous accounts of the likely consequences of some forms of decision rule: for example, in the worlds of OR and computing there have been many occasions where recommendations to introduce some form of abstract technology for production control have on implementation been a graveyard of ambitions, with the flow of work not improved and if anything made worse. Such erroneous accounts derive in part not simply from misunderstanding a production system but also from misunderstanding the content of the wider action programme of which the act of production forms only a part.

In any case there are many situations where the actors rely on limited articulation whose extent is customary. The commercial world is saturated with proposals about what is to be articulated and what is to be done if the proposals are not acted upon: for example, stock-exchange regulations on insider dealing, laws about the point of illiquidity where trading should cease, codes of truthfulness in advertising, representations of fitness for retail sale, and the articulation of the positive virtues of a product with such reticence about its drawbacks as is sanctioned by law and custom. There are other situations where the actors rely on controlled partial articulation of what they have in mind, and where the open construction of joint understanding is foreign to the intentions of the actors: it would be unusual for union and management representatives to articulate all that they significantly had in mind at the beginning of negotiations on wages or working conditions. Articulate intervention in those contexts would have to be on behalf of one group; general articulation would be unacceptable.

Yet language is undoubtedly an important element in the success of our species, and in some areas clear and well-criticized articulation is crucial for development: that seems to be overwhelmingly the case for the advanced artefacts our civilization produces and for the extensive institutions that shape our societies (provided you accept the proposal that we should regard our technologically advanced societies as an improvement on earlier ones!). The case for articulate reflection before action is not therefore that it always gives improvements but that it does so often enough to be a major source of advance, and in default of similar innovations elsewhere to be an important factor in the natural selection of action programmes.

The case for articulate intervention is similar: there is sufficient evidence of articulate intervention causing pivotal improvements for it to promise to be a significant source of long-run improvement. However, the kind of articulate interventions I have in mind is rather removed from some of the brasher examples of the peddling of abstract technology, whose embarrassing nature for the rest of the OR community lies not in the often fine technology but in the limited understanding with which it is offered.

Some of the confusion lies in regarding the abstract models of OR in rather the same way as one might regard the abstract models of mathematical physics—as ways of making statements that are close to the truth for a wide class of real situations. But the abstract models of the OR world have not at any stage been proven as appropriate models for identifiable classes of real situations. Their use must be freshly devised, criticized, and modified for each particular application. Despite the virtually absolute precision within abstract technologies, locating them in the goings-on of each new situation is a professional craft in its early stages, and criticism of the abstract model in relation to the real situation is an ongoing activity rather than something settled beforehand. That may, of course, be a reasonable way of working: an example of the sort of interplay between action and deliberative argument that characterizes all human activity. Yet it is an activity which seems to be undertaken with a weak sense of the complexity which should be addressed if the would-be-scientific posture is to be consistently maintained. Again, that is understandable. In the wartime activities I described in Chapter 1, most revisions of understanding could lead to actions which could be checked-out against the real system within a matter of days or weeks; there was the inherent possibility of interplay between abstract understanding and real system at the traditional rate of experimental and laboratory science. For good or ill, the OR communities of today find themselves attempting enquiries into systems that do not yet exist or attempting elaborate means of altering present systems where it is impractical for the elaborate means to be checked beyond reasonable doubt before implementation. This would not be particularly important if the OR community were not induced and self-induced to see itself in the scientific tradition: a tradition where the interplay goes on away from the sponsor and is taken to some fairly specific conclusion before being presented. Insofar as that leads to a picture of would-be-scientific intervention as ask-a-question-and-get-an-answer, it diverts attention from the more useful notion of articulate intervention as a joint programme of intervening and entered programmes.

The other weakness of using the paradigm of natural science is that the intervener is in communication with and of essentially the same kind as his subject matter. His investigation is in principle very complicated if he attempts to be scientific to the limit; he is unlikely to, but he may make his cut-off decisions with somewhat more insight if he sees something of what could unfold if he wished.

5.6 OR as articulate intervention into articulate programmes

Even a limited picture of the implicit complexity of intervention, and a picture which does no more than reflect the complexities which are dealt with without difficulty in the course of using natural language, is surprisingly complicated. In order to convey the formal structure of the limited picture, some sort of animated film is probably necessary! I will try to convey something of what I have in mind by a stylized story and a series of diagrams. Because I want to consider the place of abstract technology, I will set the story more in that direction rather than in the direction of classical OR: classical OR is in some ways a simpler case, with the emphasis on establishing quite basic understandings rather than on progressing along a fairly well defined road to an endpoint already imagined in outline by analogy with other interventions—or is the imagination quite as reliable as it seems?

We may imagine that Anna Liszt, a member of the OR group of an oil company, has been assigned the task of giving OR support to Chez Duller, who is responsible for on-line control decisions at one of the smallest and simplest refineries. She has been told that the kind of support she should work on is some form of computer modelling that Chez will be able to call on as a routine to help him in choosing from a range of possible actions as the need arises; perhaps wisely, she has been told not to indulge for the present in dreams of devising a system which will announce that a need has arisen even though on criteria at present in use no-one would think so.

Typically she might first set out to understand the refinery and its environment (figure 1). She does so not without preconceptions, but with a strong notion of what her view might eventually be: the level of abstraction at which her core concepts exist will both help towards some definite output and restrain the degree of divergence

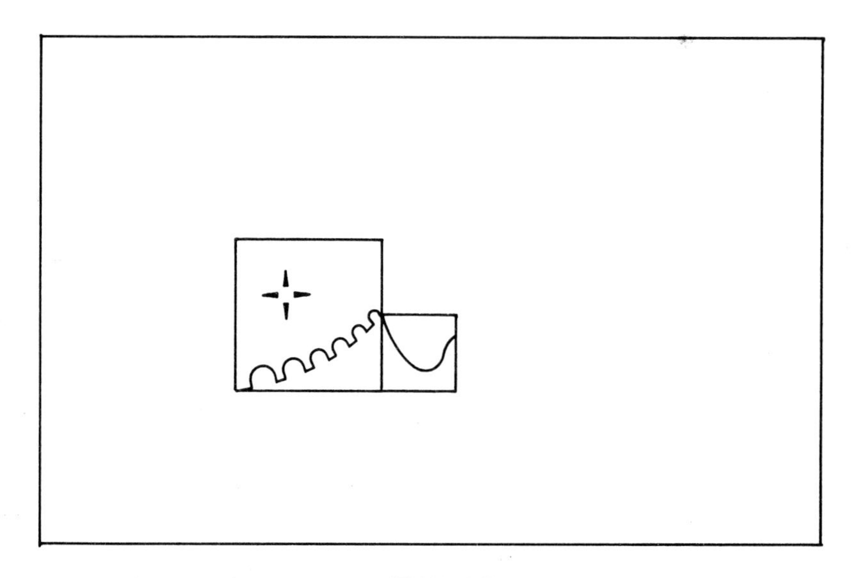

FIGURE 1 Symbolic representation of a hypothetical small and simple oil refinery and its environment—the starting point for an OR derivation of a computer model for the routine guidance of on-line control decisions.

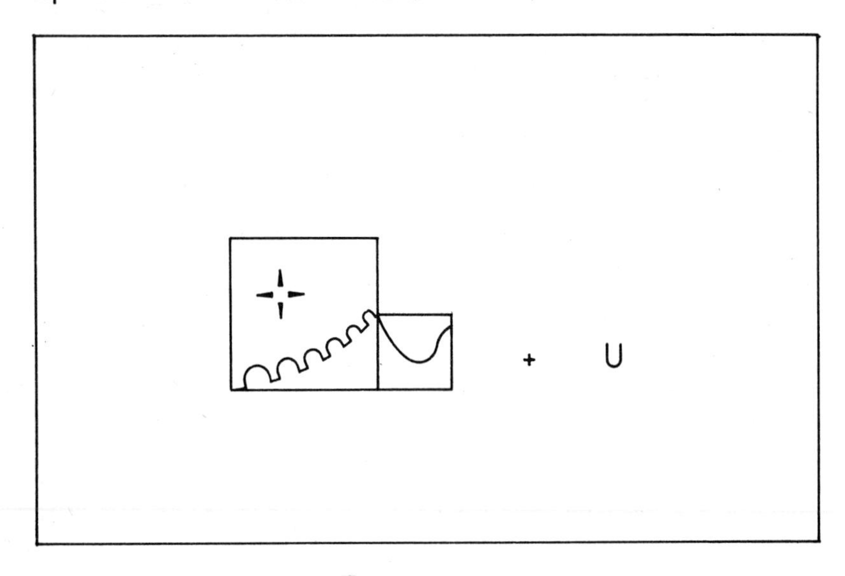

FIGURE 2 The symbolic classical end-point of a scientific study based on Figure 1 and leading to refinery control routines.

likely during the course of the intervention. If she had a classical interest in the refinery, her intention would perhaps simply be to end with some published statements of understanding about how the refinery works (figure 2): the published understanding might then be wholly or partly taken into Chez's thinking or it might soon be forgotten for lack of any deliberation in which it is rehearsed. As it is, her imagination already runs to:

(1) Getting an adequate understanding of the refinery and its environment, despite that understanding being a tremendous reduction from the present and future reality,
(2) Expressing her understanding in formally precise terms saying how, or perhaps simply that, a possible complex of actions, *x*, would lead to a complex of consequences, seq *x*, and that this would be so for a whole range, *X*, of alternative possible complexes of actions (she is wise enough to know that:

she cannot think up all the possible complexes of actions that might be undertaken, but she hopes that the range she imagines, X, will not often leave out an alternative that Chez might want to consider—that is, that the formally precise terms she uses will allow Chez's future ideas to be translated into them and will not be structurally limiting,

she cannot construct seq x to include all consequences, but she hopes she will be able to construct it in such a way that Chez does not often in the future find that $x \rightarrow$seq x significantly misrepresents the refinery to him nor that it omits the representation of consequences that prove significant to him).

(3) Realizing her understanding in the form of artefacts (e.g. papers, computer programmes, transducers for data input) so that Chez will be able in the course of his deliberation to discover seq x for any x he may specify within the limitations of X.

It is possible that her imagination also runs to:

(4) Extending her view of seq x by making statements in precise terms about Chez's ways of thinking about the merits and demerits of a particular x and its consequences—perhaps we can think of them as his open-ended bundle of criteria from which Anna hopes to formalize some specific set as crit(seq x) over the range of action complexes she hopes to consider,

(5) Realizing her understanding in the form of artefacts so that Chez might have immediate computation of crit(seq x) for any x he might consider, possibly so that Chez might be able to concentrate on just a few contenders from X that are in their various ways most interesting to him, and possibly even so that Chez might be presented with a unique member of X which seems on the understanding of actions and their

consequences and his criteria to be the most likely to be acted on by him in preference to any other—an indication of what he might like to decide to do, dec{X, crit(seq x)}.

So she starts with the picture that in several months' time she will have understood the refinery, represented it and the range of actions possible, and realized the representation and some ways of benefiting from it (figure 3). And because she is a very capable person she arrives some months later at the point she intended, or something recognizably like it (figure 4). On the way she has formalized her view of what Chez could do; does the X she has arrived at reflect how Chez thinks or has she rather made her view of what he can do fit her almost-unconsciously assumed structures that she studied so long on her graduate programme? She has also produced a number of bold assertions on $x \rightarrow$seq x which she has discussed

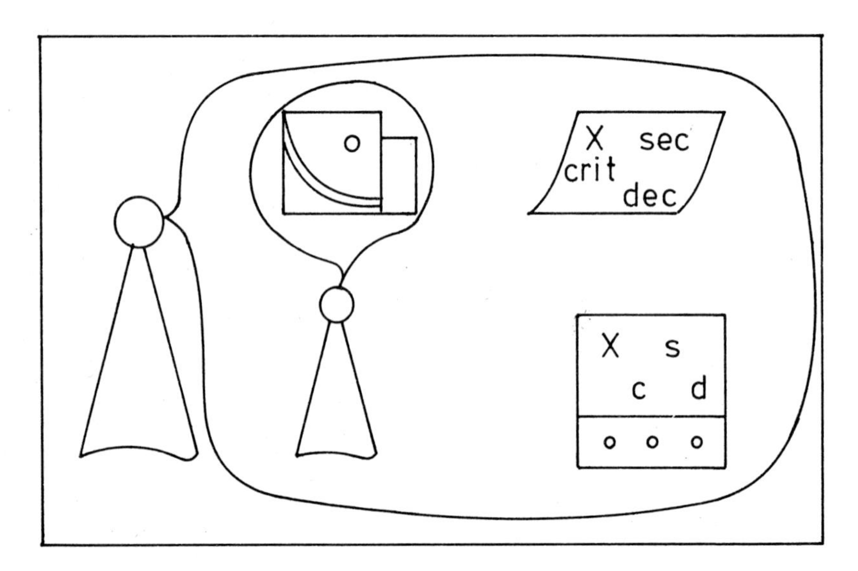

FIGURE 3 The 'thought balloon' of Anna, the refinery's OR specialist, as she sees herself expecting to reach understanding and precise representation plus a computer realization of the latter.

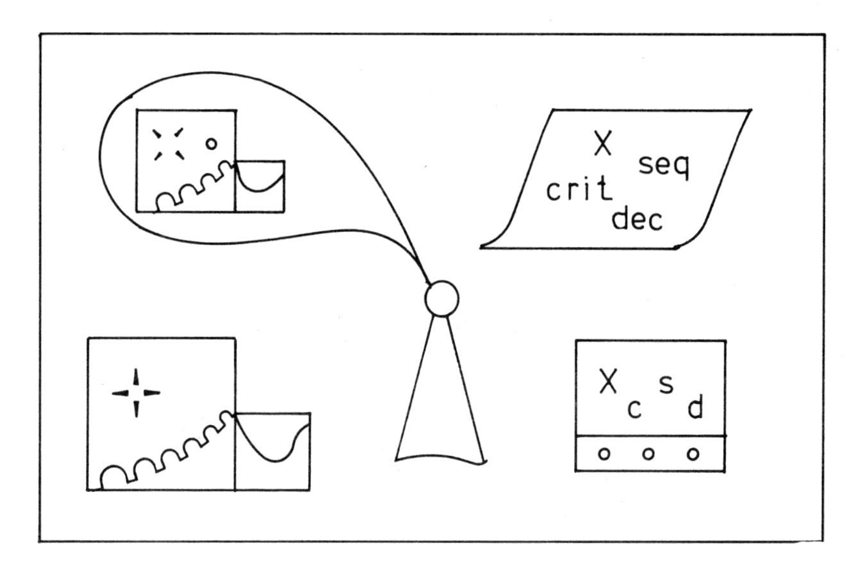

FIGURE 4 Anna's achievement after several months: a symbolic presentation of progress arising from Figure 3.

with a number of people but which remain to be proven in practice; are these going to need revision? She has also produced a number of fairly standard financial methods of evaluation, which fit some of the declared company criteria, but it remains to be seen whether these are the basis on which Chez will act, so she is already firming-up her view that for the present her intended use of optimization methods for dec{X, crit(seq x)} had better be thought of as indicative rather than normative, as something to be proven rather than as right by virtue of the strict logic *within* the model.

At this point she could if she wished declare that the project is in principle complete, and that the rest of the work is no more than implementation and the tidying up of loose ends. If she did she would be doing no more than has often been done in OR projects. It would, however, be a fairly limited understanding and one which could be extended in various directions.

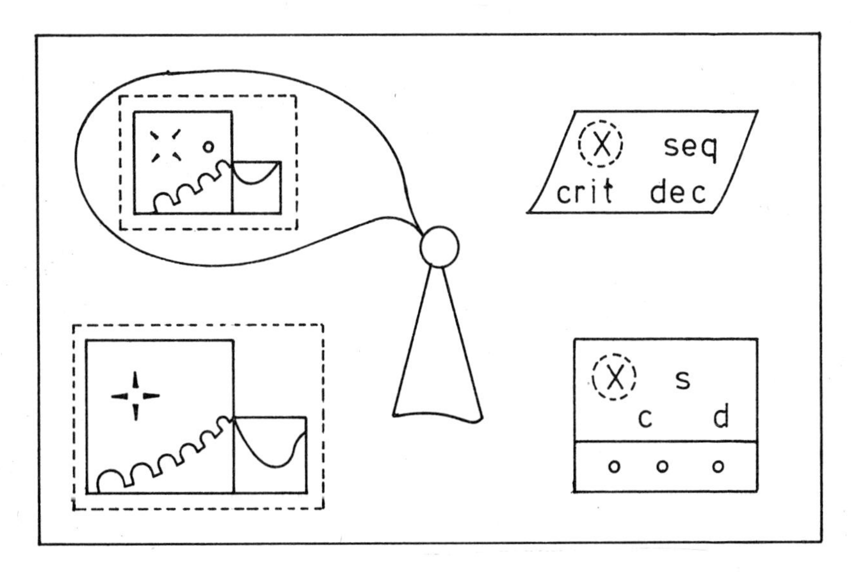

FIGURE 5 A minimal data-oriented view of change: as Figure 4 but with broken-line boxing signifying the inevitable changing of refinery variables with time.

For a start, the refinery changes through time; in figure 5 this is indicated by encircling with a dotted line. It implies at very least that Anna should see it as changing, and that some of the specification of the range of possible action will need to be changed. This in turn means that Anna might usefully consider the flow of characterizing information into the computer realization. She will at least need to make provision for that flow if her work is to remain useful to Chez. If she were to hold to a scientific ideal she would at that point need to go much further and to construct a model of the various consequences of alternative means of collecting information and of their different impacts on the refinery, Chez's programme, and on her own programme, and would at all stages of the construction have to check-out her ideas by observation and experiment: that is, she would have to embark on a long subsidiary study. In practice she will theorize about the outcome of such a subsidiary study and embark on it or not against a background of conflicting theories and proposals from the various actors just out of sight of the main action of the story.

Such a study would be of limited value since although day-to-day changes might easily be regarded as changes in the numbers held in the perhaps very extensive description of X, it would be likely that there would be a succession of changes in the refinery which would lead to changes in the structure for describing X and for describing $x \rightarrow \text{seq } x$. Moreover, with experience of using the realization, Chez might well articulate new criteria which implicitly require Anna to amend 'crit' or 'dec', or he might complain of the awkwardness of accessing the realization, or of waiting for it to respond, which would imply acting to improve the inner content of the realization quite apart from its specification. And in any case, even if the refinery and its environment did not change there would still be the

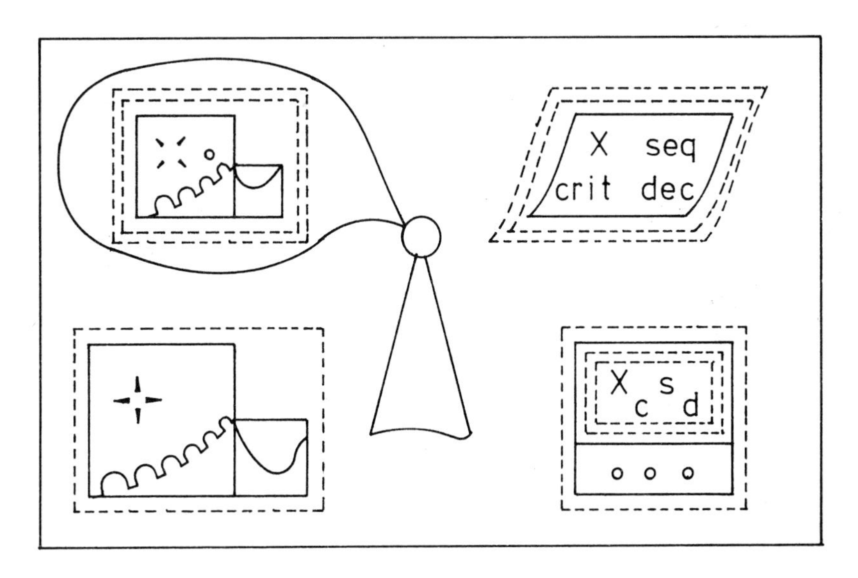

FIGURE 6 Changing views about a changing world: as Figure 5 but with additional broken-line boxing or encircling signifying the changing of Anna's own approach, arising from new criteria articulated by Chez, the refinery controller.

possibility that the use of the realization or further enquiry would lead to changes in understanding. So we can regard Anna's understanding as doubly changing: a changing view of a changing system (figure 6). One implication of this is that Anna will need to make provision for continuing modifications to be made to the computer realization either by herself through a continuing link with Chez, or by someone to be trained locally, or by Chez himself. How easy modifications are will then depend critically on the internal design of the computer realization: for example, it will be easier if the realization is written in fully modular form, if some modules deal with the interface with Chez in his language, if other modules are sufficiently generally structured to be able to accept quite major additions and deletions through the interface module, and if the operating hardware and software are convenient. The theme of designing for change is an important one for the computer and OR worlds at their interface.

But the picture of the modelling process so far outlined would serve just as well for Anna studying an automatically controlled process: Chez does not appear in it—it is simply confined to the refinery and the models of the refinery. If we try to represent the situation as it might appear to us if we were given some magic insight, we shall like everyone else be able to see the tangible changing refinery, the tangible changing precisely formulated model and its changing computer realization, and the tangible Chez and Anna. We might in addition see that Anna has by now a rather complex view in which she not only is able to conceive the changing refinery and the representations of it but is able to reflect about herself reflecting about it. She is also inclined to credit Chez with the same self-reflectiveness (figure 7). In that last point she is mistaken. Chez knows his job and he sees what he sees: a changing refinery, a girl who perhaps to her credit goes on changing her mind, and a computer system that also keeps on changing and looks like continuing to do so. Not for Chez the luxury of self-referring thought unless he happens to be of an unusually philosophical turn of mind. But for Anna it is different: she is now at a level where she can consider the effectiveness of her intervention practice in enhancing her ability to provide Chez with improved representations. Is she

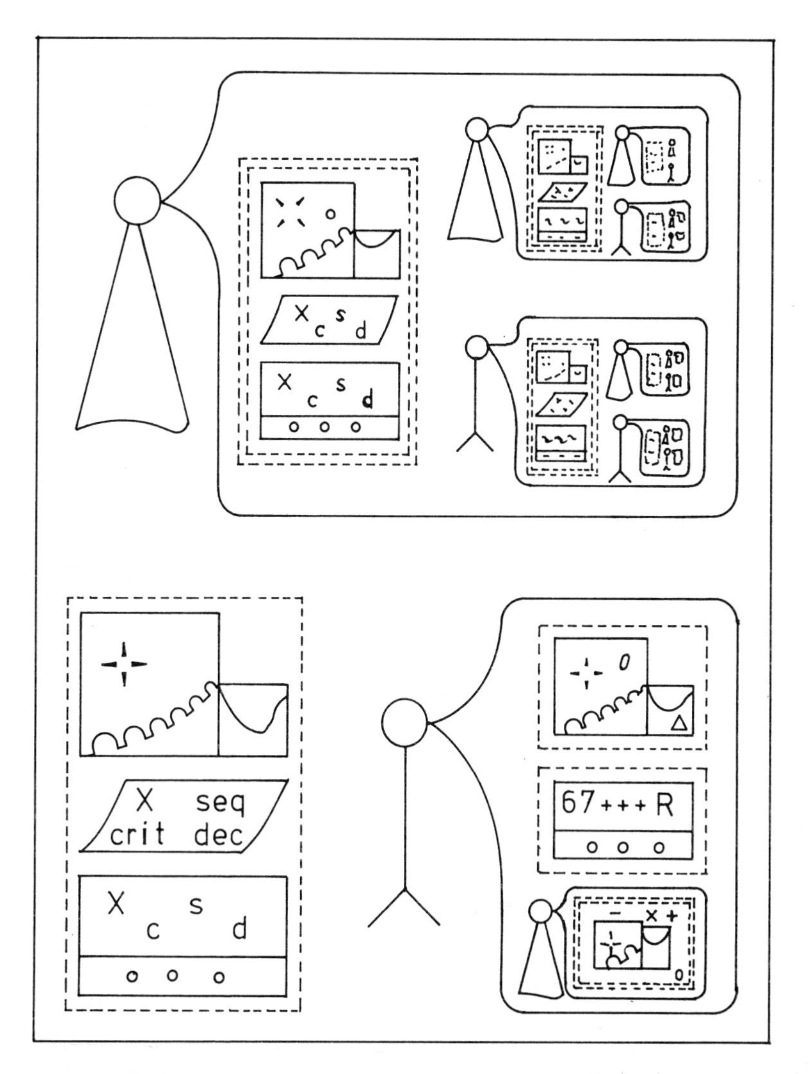

FIGURE 7 Reflexive reflection in an on-going intervention: the ever-changing oil refinery and its environment plus the 'thought balloons' of OR specialist Anna and refinery-controller Chez as they reflect upon their progress towards the desired computer model for guiding on-line control routines.

effective? Has she even begun to understand how to tap Chez's insights? Compared with other present-day interveners, probably yes. Compared with what is waiting to be discovered and invented, almost certainly not.

6. Theories

In this chapter I rehearse the status of theories, make some consequent points about the implication of their status for action programmes, and comment briefly on the treatment of theories of theories in two research programmes, one concerned with enquiry leading to scientific understanding and one with enquiry leading to understanding and action.

6.1 The status of theories

In looking at the structure of intervention we have seen that in an action programme many theories are likely to be not particularly well founded. Nor is there any chance that all the theories in a programme can be checked-out. So would-be-scientific intervention is at best a way of getting things right*er*, not of getting them right. It is a tribute to the progress of our species that what we do after theorizing can in so many cases be left unchanged for years or even decades, but it would be idle to deny that much of what we do needs continual theorizing even if at relatively low levels and that we need continual recovery from our mis-theorizing. Even the finest and most widely believed scientific theory has the status of not-yet-refuted conjecture, and for various reasons every other theory we have, whether general or particular, has the status of a not-yet-refuted conjecture.

First, the truth of a theory cannot logically be claimed to have been demonstrated for any *particular* instance in which the correspondence between theory and reality has not yet been examined:

that is, where the prediction or retrodiction for that instance has not yet been checked with observable facts for that instance. This immediately implies a conjectural status for all theories about indefinitely large classes of instances and for all theories that cannot be tested. So this first point sweeps into the category of conjecture all the theories we think of as scientific theories and all theories about the future!

Secondly, the truth of a theory cannot even logically be claimed to have been demonstrated for a particular instance in which the correspondence between theory and reality *has* been examined. The act of examining an instance of correspondence between theory and reality is itself theory-laden, even if the theories are about quite humble matters of perception. What we might slackly think of as an instance of correspondence is perhaps better understood as an instance where non-correspondence was not demonstrated. This is not simply playing with words: it is an important point. It means that as critical methods develop, theories of long standing can be toppled: for example, there might be advances in instrumentation or there might be new elaborations of the deductive consequences of a theory. This implies that all theories of reality risk later refutation in instances previously thought to be non-refuting. So this second point not only confirms the conjectural status of the theories of the previous paragraph but also sweeps into the category of conjecture the remaining theories about finite classes of instances which are entirely in the past and present, including theories about unique instances and theories about history and historical events. It applies to all documentary and personal evidence.

Thirdly, since there is no possibility of a language or a theory in which there are no undefined terms, a continuing search for clarity can unearth unsuspected shifts in meaning in the course of a logical argument. This means that even intendedly abstract arguments about abstract systems have the status of conjecture. That applies to all logic, including mathematics.

It follows that all our use of theories is provisional, and that we attend to the conjectural status of some theories and ignore for the present the conjectural status of other theories as a result of theories and proposals in the particular complex of programmes we are participating in at the time.

This is by no means the same thing as asserting that all theories are equally unreliable nor as asserting that all theories are relative to each other and not to reality. Far from it. The evidence of the natural sciences research programme and of the natural-science-based engineering action programme of the last three-hundred years suggests that we now have a number of very reliable views. Of course, tomorrow *might* bring a different world in which all our knowledge proved useless and in which the consequences of our actions were wholly other than we expected; so far the small-scale features of matter and energy have shown no such discontinuity in their nature, but discontinuities of natural and social environment are common (e.g. earthquake, currency collapse, revolution) and leave who experience them with a sudden disorienting experience of theory failure.

The problem lies not in all theories being equally unreliable but in giving any plausible account to ourselves about why we see ourselves as justified in regarding some theories as being more strongly supportable than others; it has something to do with their testability, with using them in varied situations, and with the advocacy of other people who are significant to us, but it leaves a mystery about the sense in which we can see ourselves as justified in feeling justified! That is a difficult enough problem for even the most promising areas like the theories of natural science or the theories of immediate sense data. For an unpromising area like theorizing about complicated human organizations the problem is far greater, since in part at any one time the truth about a human organization is the truth we have made for ourselves in the meanings we have ascribed to it.

The conjectural status of all theories may seem surprising, and their unity in conjectural status may appear to provide little comfort! Paradoxically it can, since it reminds us to leave all our theories permanently open to criticism, and thereby provides a key idea for a view of the growth of understanding and capability which is not manifestly at odds with historical evidence and with the passing evidence of our own experience. By contrast, the view that supposes that we have a stock of true knowledge and correct skills, to which we steadily add, neither accords with the development of children and adults we can see around us nor accords with the development of any field of understanding seen as a programme through time.

Most of this present section is freely summarized from the work of Karl Popper, and many of the condensed remarks have been the subject of careful discussion and extensive application by him. In places I have pushed my summary to conclusions suggested from Magee's account of his work and from the work of Lakatos. I first read Popper's view on the conjectural status of knowledge at a critical point in the management of OR teams when I was concerned with the logical status of OR at a time when sponsors and interveners seemed to conspire to present OR as a source of certainty, which it most clearly was not from the evidence under eveyone's nose. Such is the grip of myth! Curiously, it was the only idea from *The Logic of Scientific Discovery* that I was conscious of retaining and using until, several years later, I returned to his work in the university context of seeking to understand the sometimes major lacunae and misconceptions of staff and students about the intellectual content of OR: a context in which the familiar self-explanation of the practitioner seemed inadequate. It was at this point that Popper's writings made a substantial impact on my core and near-core theories, and provided the basis for my later appreciation of Lakatos and then my subsequent divergence from them in my wish to say useful things about the activities of the OR communities at the meeting point of science and social affairs.

6.2 Theories in action programmes

Ideally we would like the theories in a programme to be true: that is, we would like a theory, in all instances for which its applicability is claimed, to correspond exactly with those aspects of reality covered by that theory. We have seen, however, that even if one or more of the theories were true, no-one is in a position to prove it.

It is arguable that I have overstated what we require of theories: that what we would settle for are theories known to be almost true in almost all instances, so allowing for some roughness in our understanding and for occasional discontinuities. For example, it seems that in OR we usually work with theories of no great explanatory depth, but which in retrospect seem to have been temporarily not too far from the truth. If we adopt this kind of instrumental view of what we require from our theories, then the force of some of the arguments about the conjectural status of theories is weakened. But the first argument about the conjectural status of theories about the future is undiminished. An emphasis on instrumentality therefore leaves the main point undisturbed, particularly if we remember that the future course of programmes is inherently not fully predictable.

In any action programme, therefore, the bundle of theories has to be used without having been proved to be true. There is no action programme which can justifiably claim to provide certainty. Abandoning the claim to certainty is one of the key steps in moving towards a description of human affairs that is not patently false. It leaves behind the magical belief that somewhere we can find a programme that can provide us with true theories and correct proposals. It pushes participants in sponsoring, entered, and intervening programmes away from the expectation of certainty toward an expectation of permanent conjecture, enquiry, review, and innovation—always provided we can bear the emotional exposure!

6.3 The place of two research programmes with theories about theories

If you are familiar with Popper's work you will know that he drew a sharp distinction between theories which are potentially falsifiable by observation, perhaps ingeniously executed, and theories which appear to be completely free from the risk of falsification. In the first category he put the theories of natural science for which well-developed practices of critical experimentation already existed. In the second category he put theories whose proponents always saw them reinforced by events: favourable political events always showed progress towards the Marxist millenium and unfavourable events always showed an interruption in the progress towards the Marxist millenium but no event be conceived that could cause the theory of progress towards the Marxist millenium to be refuted!

Later he felt the need for a more general category than falsifiable, or perhaps for a middle ground: the criticizable. All three categories have their place in thinking about articulate intervention in action programmes. The falsifiable carries with it the notion of the hard-edged findings of ideas left standing after imaginative observation which leaves no room for serious doubt according to the standing proposals of the scientific community of the day: it includes general theories which have been put to the test, and particular observations where the chain of argument is not a matter of controversy—imagining and putting to the test falsifiable theories is a classical picture of scientic intervention to produce well-established discoveries. The non-falsifiable carries with it not simply the untestable ground theories of being and thinking that we use, but many of the notions by which we organize ourselves: theories of long-term consequences of various ethical principles and of various forms of social organization. The criticizable carries with it the notion that theories can be appraised against observation provided that one can start with a declared point of view which is one of many, but from which it is possible then to declare that, from that point of view, some existing theories are incompatible with each other or with observation, or to declare that from that point of view one can discern coherence over a wider range than has been noticed from other points of view.

Perhaps we could say that the would-be-scientific intervener typically would like to work with the falsifiable category, practises mostly in the criticizable category, and largely goes along with the non-falsifiable category without discerning it as such. But that can only be a rough idea: the boundaries are not sharp, and an imaginative approach can shift theories towards the falsifiable.

I thought at one stage that the Popper–Lakatos programme would provide most of what we need for considering what we mean when we think of an OR intervention producing increasingly reliable theories and proposals over time. But by setting OR in the context of indefinitely modifiable programmes, I have created a gap with their accounts. Popper argued that explanatory power would be increased if a theory extended the testable consequences and that the degree of corroboration would be increased if a theory passed some purposely severe test designed to contrast with earlier tests; it is worth pondering his idea that passing the same test many times adds nothing to corroboration, an idea that might hold in check some otherwise plausible arguments. Lakatos took the growth of reliable scientific theories much further, and in doing so showed that a number of concepts had been insufficiently distinguished, but in doing this he sharpened the contrast between reflection in particular situations and reflection aimed at approximating truths about general classes of real-world features—it was the latter that were his concern. However, his work on the heuristic power of research programmes is relevant to intervention programmes that include the proposal that, over a series of interventions, there shall be addition to general understanding: that would seem to be true of university research, in-house fields of work with related content, and external consultants who aim to develop increasing expertise in some area. He introduces the idea of the heuristic power of a research programme: that its core not only suggests novel observations of a kind not previously made, but also suggests novel supporting theories in advance of specific inconsistencies as the field of application is extended. By contrast, there is little heuristic power in trial and error. And in perhaps even greater contrast is the research programme whose core theories are constantly being tampered with to weaken their range of application as new counter-examples are discovered: a degenerate research programme.

The idea of heuristic seems to me to be generalizable to a notion of heuristic power not simply for truth-seeking research programmes but for action programmes in general. We would then see heuristic power in those programmes whose core theories and proposals appear not simply to be undisturbed but additionally to prompt a whole range of new supporting theories and proposals, partly in advance of the need for them.

In those terms the Churchman–Ackoff programme is one of undoubted heuristic power which, in addition to containing a research programme of enquiry into the nature of action-oriented enquiry, also has a considerable related programme of imaginative specific interventions. The interventions are intended to take new forms for new aspects of programmes in new arenas, and to avoid the simple repetition of the main features of any previous intervention. In the OR world it is, by a large margin, the most extensive programme of enquiry into innovative enquiry. Churchman's *Prediction and Optimal Decision* and *The Design of Enquiring Systems*, Ackoff's *Scientific Method* and his joint work with Emery *On Purposeful Systems* are some of the peaks in an extensive output which has stimulated many associates. It is a programme which is rooted in the North American philosophy programme of pragmatism by virtue of the early years of both its leading participants. This means that is it relatively free of the grosser misunderstandings of science from which the rest of the OR world frees itself only with difficulty: misunderstandings which are renewed with each new cohort of entrants from the world's educational programmes! By the same token it is free of the need to rehearse solutions to some of the basic problems of knowledge and action taken to have been resolved by earlier philosophers at various points in the past 200 years or so. This opens up a gap between it and the rest of the OR world. Indeed, the philosophical flounderings of others can be a source of some surprise and amusement. For example, there was in 1971 a pained reappraisal of the professional practice of OR in *Operations Research* which in parts overstated what science can achieve and, in so doing, unintentionally demonstrated the somewhat fragmented basis from which many OR interventions are conducted—this elicited in 1972 a boisterous collection of comments from the Churchman–Ackoff programme in *Management Science.*

From various clues I can sense something of the significance to the Churchman–Ackoff programme of people like Singer, Moore, Peirce, and some of the standard European 'greats'. It seems unlikely that other members of the OR world will get much benefit from reading what they wrote; their presuppositions and preoccupations are in varying degrees remote from our situation, their significance to the Churchman–Ackoff programme lies not so much in what they wrote as in the significance with which their writing was regarded in the North American philosophy programme of the second and third quarter of this century, and they are open to quite basic criticism from the point of view of what is now understood about language. What does come strongly through the years is their courage to think.

These are some of the structural reasons why the Churchman–Ackoff programme has not long since become the perceived adequate ground for OR intervention, even though it has considerable heuristic power in suggesting interventions to undertake, in dealing with reverses within them, and in prompting enquiry into the processes of enquiry and action. What seems instead to be needed is a directly accessible rehearsal of basic theories about the place of theory-making in other people's goings-on: a rehearsal which appeals directly to experience of self-conscious deliberative argument rather than arriving there on other grounds. This comes over to me so much more strongly in the Popper–Lakatos programme that it seems to me to be the programme to work from for the OR community I know at first hand.

I suspect my choice is culture-influenced. In contrast with the US, organized life in the UK is I suspect less mistake-acknowledging, so a healthy awareness of the conjectural status of theories is long overdue.

7. Testing Theories

After a fuller statement of some of the problems of theory-checking that have been mentioned at earlier stages, this chapter moves on to three aspects of theory checking that are relevant to would-be-scientific intervention: the mass of implied theory re-appraisal that is already explicit or implicit in a programme to be entered, some of the particular problems of checking the arguments supporting an abstract technology, and some comments on the arrangements for criticizing the correspondence between some abstract technology and some particular goings-on. This leads on to the view that OR interventions, and perhaps other forms of would-be-scientific intervention, must be seen as a wider form of critical participation than the scientific. The final section discusses some of the hopes with which statistical methods might be invested, particularly by those whose main approach to them is by way of theorizing about them: it is written primarily for those who already have considerable technical expertise and are (or ought to be) troubled to know how to regard the relation of that expertise to real systems.

7.1 The problematic nature of refutation

At first sight it may appear that it would in principle be a relatively clear-cut matter to check the correspondence between theory and reality at some given point in time. For at that point in time we would have the currently non-refuted theories of how to check correspondence for that particular kind of theory, and so we might expect that after a series of actions we would arrive at the point where we could say that the theory had become refuted or that it had survived unrefuted. For example, in an audit check we might expect to reach the stage where we can declare that in the previous year there has been no systematic fraud.

There are however various reasons why refutation can be problematic. Of these, three have particular significance for OR interventions, which have links with real goings-on on the one hand and links with the world of formal abstract technologies on the other hand.

First, the meaning of an observation depends on a network of supporting theories in addition to the one under scrutiny. So apparent non-correspondence brings all the supporting theories under suspicion in addition to the one under scrutiny. For *each* of the supporting theories there is therefore a new subsidiary problem: that of devising and executing a test to check whether that is the source of the apparent non-correspondence. But each of these subsidiary tests itself depends on its own subsidiary network of supporting theories, which may in turn fall under suspicion. So an apparent refutation produces a potentially infinite cascade of problems to solve.

The new problem of having an infinite cascade has no logical solution. In practice the problem is resolved by partly consciously and partly unconsciously not going too far into it. Each action programme has its own stopping behaviour linked more or less articulately to its core proposals and core theories and influenced by proposals from other programmes. For example, it may be asserted that some particular subsidiary problem has a trivially obvious solution (e.g. that a particular transaction took place because we have in our hand a cheque covered with all the usual signs of a document that has passed through the banking system) or it may be asserted that there is no means of testing some other subsidiary problem (e.g. whether the signed document over a dead industrialist's signature is an expression of his intent). It is more likely in deliberative argument, however, that the cascade of subsidiary problems will dry up in an unchallenged use of language whose latent and implicit content is not consciously evoked, or at most that it will dry up in a general assent that there are better things to do with one's time. In a more formal setting there are notions of the limits to which citizens should go in exercising due care and attention in their professional roles: how far an engineer should go in examining a structure for safety, what constitutes reasonable pre-flight checks by aircrew, what degree of probing is reasonable before making a social security payment, and the extent to which people in their professional roles can be inspected in order to assess the adequacy with which they carry out the theory-checking at that time considered the norm for that profession.

There is in the OR world no escape from this essential feature of the human condition! You will terminate your enquiries consciously or unconsciously at some logically incomplete point. You cannot get to a point of logical finality for yourself nor can you truthfully offer it to anyone else.

Secondly, the symbols with which we conduct arguments are to a greater or lesser extent fuzzy, in the sense that they induce somewhat different groupings of images in different minds. This opens up the possibility that in one mind there is assent to the claim that a case has been logically made, whilst in another mind there is dissent from the claim. Since we have no absolute authority to which to turn to ask whether a case has been made, we have to acknowledge that we may have to live with dissent, that because of the fuzzy nature of their meaning a string of symbols representing a logical argument does not carry within itself any means of endorsing the correctness of the logic, and that the correctness of an argument is a matter of debate. Even quite short logical arguments can fail to win assent. For example, Popper's work is full of footnotes recording battles over whether he has or has not provided a logical demonstration, particularly in the field of the foundations of probability theory and the inappropriateness of probability as a measure of corroboration. Where arguments are complex, and where the symbols point to social concepts, there is almost infinite scope for disagreement. The OR world might usefully regard the clarification of meaning as a standard activity.

Thirdly, the theory tested might be in the form of a probabilistic model, in which case any set of observations which are conceivable under the theory are logically non-refuting, however improbable is the set under the theory. In many action programmes this is dealt with by introducing the notion of almost certain refutation. The notion of almost certain refutation is formalized in various ways in various approaches to statistical inference (e.g. by the rejection of a hypothesis, or by the quotation of a low likelihood ratio). Within the programme of research into the principles of drawing inferences from the imagined output of an abstract probability mechanism, there are well-understood conventions about which parts of the cascade of supporting theories will in imagination be subject to further examination: the conventions change slowly with time and reflect in a rough way the problems that arise in the use-of-probability-and-statistical-theory action programme. The two programmes are, however, quite markedly different.

The problematic nature of refutation does not mean that we need to give up the idea of theories being not-yet-refuted conjectures, simply that refutation itself remains uncertain. Since theories are differently regarded in different action programmes, it will not be surprising if a theory is regarded as overthrown in one programme and not in another. In any case, even if an existing theory is understood to be overthrown in principle, it may well continue in use with an awareness of its falsity until such time as there is something better to replace it. Moreover, if the falsity is seen not to be so much an error of category as some small error in the magnitude of some measure, then the theory may continue to be used indefinitely as an approximation to a truer one. That may be in the absence of a closer theory or it may be as a conscious convenience or it may be in ignorance of a closer theory already available in some other programme.

7.2 The review of theories

In view of the provisional and conjectural nature of theories, it might be thought that action programmes would normally contain proposals for the active review of their theory content. Apart from programmes one of whose core proposals is that their theories should be actively reviewed, research programmes for example, it seems that there may be a tendency for action programmes to cluster towards the other extreme of reviewing theories only when some inconsistency obtrudes. This can be an alarmingly weak approach since, by the time some inconsistency is apparent in a regularly used theory, many opportunities may have been passed of examining its network of supporting theories.

This potential weakness is partly met in industrial, commercial and financial programmes by a wide range of specific proposals in their own programmes and in the general legal programme, although the frequency of legally required reviews may well be far too low to achieve the overt purpose of the legislation. At a fairly general level we have the network of practices by which national or regional plans are formed for several years ahead, though the dissatisfaction that they cause and their renowned divergence from what then happens suggests that the processes are far from complete! At a lower level there are the devices of annual meetings of shareholders, the planning meetings of senior managers, the presentation of detailed monthly accounts, and the day-by-day computer record of sales transactions. All these are devices for amending and (hopefully) improving the theory content of action programmes.

At one end of that spectrum, the use of the word theory may seem natural: at the other end of the spectrum words like 'data' or 'information' might seem more suitable. However, the use of 'theory' to indicate what we normally think of as data was neither careless nor silly: even the supposedly hardest data came into being because a theory-laden set of people caused them to be created by a theory-laden methodology either for themselves or for a theory-laden set of others.

If the intention of would-be-scientific interveners is to leave behind at least some improvements in the entered programme, it is quite possible that they might make their contribution from the point of view of asking what the theory-making capability is within the entered programme as it stands, and what its theory-making capability might be under various alterations. There is a very important area here which should be sharply distinguished from the flow of numerical information under presumedly stationary structures of meaning. What I am concerned with is how relevant theories are to be articulated, criticized, stored, used and reviewed. If the aim of intervention is to improve the deliberative argument in some way, then the processes by which theories are accreted cannot lightly be ignored. They are as relevant a subject for study as the matters the programme is overtly concerned with.

7.3 Theory checking in abstract arguments

In would-be-scientific intervention, the range of theories met with is perhaps wider than is usual in the entered programme. It is quite likely to include at least some theorizing about the properties of an abstract system which represents some imagined reality which is therefore not immediately available for checking correspondence against.

The argument in such cases is likely to proceed by mathematics or computation or some mixture of both. On the face of it such methods appear precise, were it not well known that they produce a steady but unpredictable stream of inconsistencies, which are deemed to be mistakes if there is a general feeling that their resolution could be achieved by many people within the mathematics programme (e.g. a misstatement of the algebraic representation of the abstract system described in words or diagrams); are deemed to be errors if not usually examined behaviour is at fault (e.g. decoding errors in input equipment due to faulty mechanisms); are deemed to be problems if the resolution looks like requiring the clever use of auxiliary theories and a period of reflection (e.g. problems that are discovered by taking apparently well-argued conclusions and discovering conflicting properties for limiting cases at the boundaries of the feasible values for parameters); and are deemed to be paradoxes if it is suspected that the use of language is obscuring some distinction that has yet to be made.

Because we know that our abstract argument is intentionally precise we may not expect too much difficulty in principle when we ask whether claimed mathematical results agree with the mathematical premises. However, even with simple arguments there is a clear need to develop a second argument as independent of the first as possible. This might be no more than a simple bounding argument for magnitudes, or an argument of the form if-the-conclusion-were-true-what-else-would-follow-and-is-that-plausible? A complex probabilistic argument might well be checked-out by simulation, though if the argument is based on general forms of probability distribution it is as well not to base the check on cases known to collapse to simple results in elementary models: for example, the poisson distribution in time-dependent models or the normal distribution in models of aggregating effects. If one does use a collapsing type it is quite possible to miss the signs that the would-be-general argument about the model has somehow wrongly presumed properties that apply only in the collapsing cases.

7.4 Theory checking of abstract models for particular real systems

It is perhaps more likely that the faults of an abstract model lie not in mistakes in would-be-logical argument about its properties, but in mistakes about representing the real system it purports to model or in mistakes about understanding its possible modes of use in the deliberative arguments of the entered programme.

When an abstract system is put forward as representing a real system, and when theories about the properties of the abstract system are offered as theories of the real system, we have a situation in which the logical correctness of the abstract argument is no guarantee of the correspondence between statements about the abstract system and the unknown truth about the real system. If the theory is adequate it will include as particular cases of the general model both the theorized past of the real system and various theorized possible futures of the real system. In that case we can expect to have retrodictions about various past and present observables, which are of theory-checking value if they have not trivially been read into the model to represent its past, and we can expect to have predictions of future observables both for immediate use in deliberative argument and to check with future correspondences as they are realized and move into the past.

Since a theory depends on an indefinitely large cascade of subsidiary theories, there is no fixed way in which all possible relevant theories can be kept under review. That is not intended to be defeatist, since it opens up the never-to-be-completed challenge of imagining all members of the cascade about which there might be doubt and of devising ways of keeping them under review. Indeed it points towards a principle which is being increasingly recognized: that would-be-scientific interveners need to devise ways in which subsidiary and component theories are not lost within the total representation of a complex model, but are maintained in accessible representations whose meanings have some direct connotations within the entered programme, and whose significance and correctness are therfore a matter for direct comment. Even if one takes a purely instrumental view, there may well be programmes in which a regular examination of the substructure will give advance warning of the dwindling validity of the point of view enshrined in the cascade and enable timely and apt changes to be made.

So, for example, Harrison's models for Bayesian forecasting have a form and a computer realization which emphasize the interpretability of components. In the field of mathematical programming models, the proposal that model transparency be aimed for means a shift of emphasis from efficiency, as measured by compacting a model to fewer variables or quicker execution, to effectiveness, as measured by the immediate interpretability to the entered programme of each numerical element, if necessary by quite freely introducing explanatory variables and connecting relationships which it would be algebraically feasible to do without.

Since theory-checking for one model is in principle of indefinite extent, it is clearly impossible to give an exhaustive list of recommendations for attempting theory-checking! I suspect that if the difference between an abstract system and the corresponding real system is kept in mind, that will of itself be a powerful heuristic: reality and its representation are often confused and the capacity simultaneously to follow vigorous roles as model creator and model critic is far from well-developed. Indeed, adequate criticism is a social rather than an individual matter.

There is, however, a further point about theory-checking that should be borne in mind. Since theories about goings-on cannot be safely regarded as having been checked-out once and for all at some point in time, some arrangement is necessary for their ongoing review. That might be as simple as setting future review dates for interested actors to meet, or it might take the form of predicting and monitoring the aggregate properties of proposed control methods in ways which are sensitive to departures from the abstract model in aggregate but which cannot detect departures at the level of individual control: for example, Wagner's extended theorizing to suggest theory-sensitive controls for a particular class of inventory control methods.

7.5 The pursuit of criticism

Because the OR world has typically in its quest for would-be-scientific intervention also supposed that that was a quest for would-be-normative intervention, it is common to find that each intervention carries latent and implicit theories about the large general action programmes of our industrial civilization. These are often at such a general level as to be untestable for the purposes of the intervention, and their use is in effect little more than a proposal from one action programme to another. For example, the theory that the selection of energy policy is best carried out by examining expected prices of fuels rather than by examining the potential for adverse political and military activity, or the theory that business will in the foreseeable future be conducted in a political and economic climate not discontinuous from the present in major ways.

Between the untestable general and the testable particular there is a countless stream of theories, every one of which presents the problem of whether it is relevant to criticize it, and what kind of critical method might be brought to bear upon it.

One of the characteristics of the scientific method is a close and tightly argued check against reality. Insofar as the practice of OR has such tightly argued checks, it can without any fear of misleading anyone be claimed that it exemplifies scientific method. Some of the earlier case histories like those of Waddington and Edie do this, and are celebrated for that reason. But with the passage of time it has become customary for OR communities to address themselves to situations where much of the content of articulate reflection is beyond any evident means of checking against reality in advance of choices to be taken in the entered programme. And insofar as the practice of OR is no longer susceptible to such checks, the claim that it simply exemplifies scientific method is no longer sustainable and must be replaced by some wider view of the critical activity it contains. This takes it closer to the general critical activities of society, where the novel is more often rewarded on the basis of conversational scrutiny than on the basis of observational scrutiny. Thus there is a kind of superficial criticism in plenty but little that, at the expense of very hard thinking, is rooted in really searching tests against reality. Indeed such tests can seem more resource-consuming and less attractive than simply acting on the basis of the model as it exists after conversational scrutiny, in much the same way that everyone else's ideas are acted on!

Yet the habits of looking for formal measures, coherent structure and systematic interactions give the once-was-simply-scientific intervener a continuing role in deliberative processes and perhaps some sort of advantage in them. But it also becomes possible, where no strong tradition of criticism has been established within an OR programme, for the programme to drift into fixed patterns of what is criticized and how it is criticized: patterns which it becomes increasingly necessary to change if the programme is to have, and to be seen to have, a distinctive contribution to entered programmes. In this matter, I fear that a preoccupation with abstract technology is the opposite of what is needed, unless it is technology that is concerned not with the normative aim of optimization but with the participative aim of supporting deliberation. Perhaps that is too bald a statement: they are by no means exclusive, and the participative aim can subsume the normative aim as one possible aspect of participative comment. Indeed, the participative aim can present abstract problems of substantially greater fierceness than has hitherto been produced by the normative aim!

7.6 Criticism avoided? Some false hopes of statistical inference

Theories of how to draw inferences from the imagined output of an imagined probability mechanism have provided an immense area of research and educational activity in the past half-century. Since the language of imagined inference includes words like 'observations', it is perhaps understandable that it has not been sharply distinguished from the language of drawing inferences from the real output of real systems. Indeed, since the original motivation of the research and its normal hope is that it will help to provide inferences from the real output of real systems, it is understandable that the two are confused. Moreover, there are from time to time pieces of theory about inference-making that claim, or are thought to claim, that theorems about imagined inference-making are true about real inference-making.

The usual format for studies of imagined inference is that a class of probability mechanisms of greater or lesser generality is evoked. It is then possible to imagine that one member of the class produces a sequence of observables, and that the sequence is then to some degree made known to an imagined observer who at that point knows:

(i) Whatever features of the sequence you have imagined to be declared,
(ii) The complete precise description of the class of probability mechanisms,

and who then addresses himself to the problem of arguing towards some precisely stated retrospective view on the relative strength of the claims of various members of the class to having been the origin of the imagined observations. There are various ways in which he can be imagined to argue, using principles which appeal to different groups. Although the principles are sometimes advocated as though they were the correct way to argue about imagined inference, and about real inference, their use is a matter of current standing proposals in particular programmes. Once the class of probability mechanisms has been described precisely, and once the principles of arguing towards a precisely stated retrospective view have been precisely stated, the remaining problem is simply one of logical elaboration from the precise statements. So it becomes possible to construct a precise computable route from imagined observations to precise retrospective discriminating statements about the imagined class of probability mechanisms. That is a problem of sufficient complexity and variety to be the chief concern of research programmes in theoretical statistics.

It is the element of computability which has often seemed to offer the hope of purely deductive inference about real systems. It is a false hope. Our imagined observer is rather a dull fellow, despite our presumption that he likes precision! He has played no part in the specification of the class of probability mechanisms, either by imagining some mechanisms for himself or by debating what is to be included and excluded from the class. And he has been no more than a puppet in his choice of principle. He can hardly serve as a model of a real inference-maker. Nor can he serve as a model of people who make such theories about imagined inference, though I know they may become excitable if reminded of it!

So how can we avoid misleading ourselves about the nature of theories of statistical inference? It can perhaps be usefully done by adding the following theory to our programmes:

> *Theories of inference are incomplete without a theory of theory-making.*

This is offered not as a theory about an abstract system of inference, but as a theory about real inference. Its validity is therefore to be considered not by looking for a theorem about it, but by looking for refuting instances. It is open to refutation by anyone who can produce an example where inference is possible without a theory-maker. Good hunting!

From this point of view the clash between classical and Bayesian approaches to inference is resolvable in favour of neither. They are both examples of particular ways of addressing abstract problems and they are both incomplete. Their advocates are participants in programmes with many theories and proposals in common, but with a few proposals (masquerading as theories?) differing. Their transfer from imagined inferences to real inference has to be accompanied by action proposals which are extraneous to the theoretical constructs within the approaches, in that there is no category for them within the constructs.

In the second quarter of this century there were great strides in the classical approaches to the theory of imagined inference. A clear distinction was maintained about acceptable forms of statement in the language of probability. It was acceptable to use the language of probability about the relative risk of alternative observables from a probability mechanism. It was unacceptable to use the language of probability to describe ignorance or uncertainty in the same simple way. It was, however, acceptable to use the language of probability to make certain careful statements about the imagined class of probability mechanisms. So quite precise statements were made, for example, about the probability that some way of constructing a sub-class of the whole class of probability mechanisms would in prospect, before the imagined observations had in imagination occurred, include the particular mechanism that would by that later time have generated them. Precise statements on inferences about an imagined class of mechanisms do not, however, provide correspondingly precise statements about real inference. In real inference, no-one has inside information from which he can construct a correct class of imagined probability mechanisms from which to choose, no-one can do better than face an open set of imagined probability mechanisms some of whose members have not yet been conceived or articulated, no-one can deduce a precisely stated retrospective view on the relative strength of the claims for any sub-class of that open set, so no-one can justifiably claim to have a correct prescription for using the arguments of imagined inference for real inference. At best he might summarize what he believed has been useful practice in the past as a guide.

However, the move away from classical statistics in the third quarter of this century took place for other reasons, one of the most important of which was that the classical approach led to statements that were not in a particularly suitable form for going on to action. The third quarter of the century has seen an immense amount of research into imagined actions in the face of imagined probability mechanisms. Just as theories of imagined inferences are incomplete as models of real inference, theories of imagined choice of action are incomplete as models of deliberation before action.

The usual format for studies of the imagined choice of action is that a class of possible actions is evoked, and that precise statements are then made about the imagined consequences and their imagined significance to one or more actors—the imagined consequences may be imagined to be determined by some form of probability mechanism. It is then possible to imagine an actor confronted by:

(1) The set of actions you have imagined,
(2) The precise descriptions of consequence and significance you have imagined,

who then addresses himself to the problem of arguing towards a precisely determined choice of action. There are various ways in which he can be imagined to argue, using principles which appeal in different ways to different groups at different times: their use is a matter of the current standing proposals in particular programmes. From time to time claims are made that some principles of choice are correct for imagined action, and it is sometimes additionally supposed that theorems about imagined action are a method of proving theories about real action. Once the set of actions has been described precisely, and once the principles of arguing towards a precisely specified choice of action have been precisely stated, the remaining problem is simply one of logical elaboration from the precise statements. So it becomes possible to construct a precise computable route from imagined choices and consequences to an imagined recommendation for action. The rules for finding and following that route are what I have previously referred to as an abstract technology. The finding of computable routes is a problem of sufficient complexity to have been the chief concern of research programmes in the mathematics of OR.

It is the element of computability which has oftened seemed to offer the hope of purely deductive ways of choosing action in real systems. It is a false hope. Our imagined actor is rather a dull fellow, despite our presumption that he has an appetite for precision. He has played no part in specifying the classes of actions and consequences, either by imagining them for himself or by arguing about what is to be included and excluded from the class. He is little more than a puppet presented with what is a *fait accompli* apart from the deductive argument which he has no choice but to accede to!

So if we are to avoid misleading ourselves about the nature of theories of choosing action, we need to remind ourselves that prior to the imagined problem of choice there is the problem of assembling the statement of the problem. Perhaps that can be simply done by adding the following theory to our programme:

Theories of choosing action are incomplete without a theory of an action-imaginer and a consequence-imaginer.

The theory certainly holds good for any system of automatic response we devise, since the abstract specification or its physical realization are observable consequences of the imagination of the designer. If we focus on the likely place of inference as an intermediate stage before action, we might formulate something like:

Theories of reflectively choosing action are incomplete without a theory of an action-imaginer, a consequence-imaginer, and a theory-maker.

From that point of view, most theories of the imagined choice of action are incomplete, since their relation to real choice has to be accomplished by proposals which are extraneous to the constructs for describing imagined choice.

Which brings me finally to the Bayesian approach to inference.

If we are prepared to use figures in the interval [0, 1] not only to represent relative risks of particular outcomes from probability mechanisms, but also to represent our view of our ignorance and of our uncertainty about the relative claims of particular members of some class of probability mechansims to be the source of a sequence of observables, it is possible:

(1) To make a combined statement of risk, ignorance, and uncertainty in a format identical to the format for a statement of pure risk,
(2) To specify computable routes from a new observation to a revised statement of risk, ignorance, and uncertainty,
(3) In the computable route from imagined choices and consequences to imagined recommendations, to replace any statement of risk by a similar form of statement to include risk, ignorance and uncertainty.

This then provides a computable scheme for assimilating observations and for recommending a sequence of actions as the imagined system develops through time.

For the reasons already given, these attractive properties of the imagined inference and action scheme do not provide a complete model for real inference and action. The precision of computation, once a model has been assembled, carries with it no corresponding certainty about the construction of the model either at the time it is devised or at any later stage. The construction of sets of imagined actions, imagined consequences, theories and proposals on which the computational routes are based, is left largely uncriticized by the computational procedures adopted. Ongoing criticism is external to the scheme. Indeed Lakatos has used Carnap's attempts to find a computable basis for inference, a theory of inductive logic, as an example of a degenerate research programme in which the core theories have been steadily whittled away by adverse instances.

So a Bayesian framework of ideas is insufficient as a model either of inference or of action. We can, however, quite reasonably adopt the proposal to use a Bayesian scheme of inference or action for some real choices where we do not expect to wish to modify our stated repertoire of actions or our theories or our proposals for some reasonable time ahead. We may already be satisfied that our model is adequately close to reality, and we may be prepared to act under the theory that reality is not for some reasonable time going to change to make our theories less suitable. Indeed it is possible to develop ingenious and powerful schemes of Bayesian response, and to find that they serve us well for a while without modification. My point here is not to decry them, but to warn against the view that we can in principle hope to free such schemes from the need for continuing review and from the need for the articulation of new concepts and new theories.

8. Symbols, images, concepts, and theories

Theories and proposals do not arrive fully fledged in neat grammatical form. The neat form is a final articulation of what has not had quite that announceable form earlier, and which we hope may be clear enough and sharply defined enough to bear the weight of argument. So we need to pay some attention to the emergence of theories and to the emergence of the concepts out of which the theories are constructed.

To meet that aim, several very different Chapter 8s could be written, and each type of content would have its own perhaps passionate advocates. It could be a chapter rooted in the neurophysiology of perception, though I am not at all sure how far that has progressed towards giving an intervener useful views of the formation of what we think of as high-level concepts. It could be a chapter rooted in some framework of understanding of the sociology of knowledge. It could be a chapter rooted in an extended theory of the linguistic structures by means of which we construe the meanings of particular strings of words. It is in fact a chapter in which I use a few notions which you might find useful for describing yourself, and by means of which you might widen your view of what you could do in a programme of articulate intervention.

8.1 Where do you begin?

If you are reading this, it is almost certainly at least 20 years too late to consider where to begin as a general problem with a general answer. You are now deep into your unique life-experience of impression and emotions and images and linguistic practice. It is quite impossible to begin again. Any intervention you undertake will therefore be accepted and embarked on from the middle of your experience, and you will be dealing with others who are deep into their own unique linguistic experience.

Indeed, articulate intervention is only possible *because* you and the others are already deep into those individual linguistic experiences. As Foucault reflected, the practice of language depends on actors leaving *unsaid* almost everything that *could* be said. He had in mind the use of natural language, but the same is equally true when we use other symbols for communication, be they algebraic statements, diagrams, accounts, traffic lights, or indicating instruments. So even the largest and most coherent formal models that are constructed represent a tremendous reduction from all that could be represented.

So where do we begin our representation for a particular intervention? Again, it is too late. In whatever moment we became aware of it, our representation not only began with whatever images we explicitly had of it, but those images, being in mostly familiar form, carried with them a great mass of associations from our lifetime of construing the world around us. Nor is there any way of starting with an almost empty memory next time!

If from school and university programmes we have drifted into the idea that problems, particularly scientific problems, have some clear beginning to begin at, some right point from which orderly argument should flow, then we are likely to imagine that for the purpose of articulate intervention we can somehow jump outside the human condition: that for real problems we can somehow get into the same godlike position that we assume over the closed abstract problems we create for ourselves. We are in the middle, and we can only proceed from where we are.

How we in fact proceed is a matter of memory, imagination, and the stream of perception experience we have organized for ourselves. The perception experience we organize provides a massive stream of visual, auditory, and other kinds of stimuli only a small fraction of which causes conscious awareness, despite conscious awareness itself having a massive richness. Again, for the purpose of intervention we may remark on and retain only a small fraction of what we are aware of to contribute to what we eventually take to be a significant representation for the entered programme. We may not be aware of all the implied freedom in that description of ourselves, since we may somewhat unreflectively accept the constraints provided by the view we have of our role and the specific proposals of the entered programme for what should take our attention. But the freedom is there.

What is retained and ordered from our massive perception experience is not altogether under our conscious control: it will in part be ordered by our already extensive pattern of mental activities and constructs, much of which is barely accessible to inspection by us. On the other hand, we are not slaves to the perception experience. We can and do exercise some choice about to what we pay attention, and do so with purpose and the intention of understanding. So we can and do in part design the perception experience

But it seems inappropriate to regard that designed experience as something for which there is a formula to determine what it should be for our purpose, and to suppose that thereafter we make unavoidable progress towards appropriate conclusions. We are not necessarily driven to any determined conclusion: we are not, to use Churchman's somewhat playful description of one of his categories of enquiry, Lockeian enquirers. Rather, the perception experience, together with our present memory and imagination, provides an heuristic environment within which further understanding and articulation and imagination can occur. The drift of this book is in part that I expect its content to improve the heuristic environment for your interventions.

But the material in this section also points to other conclusions, of which I will select just two that seem particularly relevant. First, the participants in an entered programme all have their on-going perception experiences; one possible result of our intervention is that we re-design the structure for the perception experiences of the participants as a revised heuristic environment for them. We are then in the business of designing information flows not against our view of how deliberative argument *should* take place in the entered programme, but against a view of how the content of deliberative argument might be enhanced. So the flow of perceptions we might engineer are not necessarily of the kind that is encoded into natural language or the symbolic and numerical languages of accounting reports. Simple examples are the selective transmission from television monitors, or the recommendation that there should be a regular programme of first-hand contact with situations that managers otherwise only know by encoded data—China's programme of massive industrial and agricultural experience for people whose normal roles are elsewhere is perhaps the extreme example! There is also for the would-be-scientific intervener here a potentially rich area of development in considering the relation of fully articulated models to the ongoing perception stream of participants in the entered programme.

The other conclusion from the material in this section is that the enhancement of intervention ability is unlikely to be best accomplished simply by articulating statements about it. There is much that happens in theory-making activity that is not captured in statements about it. Indeed, statements probably have greater use for the organizing of understanding about experience after the experience rather than as a precursor of it. This has all kinds of implications for education for intervention. It more than justifies the use of case histories, case studies, and participative project experience, but it also suggests a point of view from which the moment-by-moment content of educational programmes might be reviewed, and the place of those programmes in the career of an intervener. It also suggests that the design of abstract technology for a deliberative context may be quite seriously deficient if the designers are without first-hand experience of that deliberative context.

All of which means that I am reluctant to suppose that we can write a manual of intervention that can be put into practice by any intelligent layman: a sort of algorithm of intervention. Naturally, as from experience and conjecture we perceive similiarities between different interventions, we build up a received view of how to go about things, which if there are sufficient people busy with that sort of intervention can for a while become the customary practice. So for example it is common for firms of management consultants to build up a house-style for their interventions and to transmit that by written instructions and training programmes. In that sense it may be possible to build up a methodology for a particular class of recurring intervention. But there is nothing in the process by which the methodology is articulated that can lead to a justifiable claim that its adoption in a particular instance will either be free of gross omissions or will be at all close to the effectiveness of what might be achieved by some more appropriate method of intervention as yet to be articulated. Yesterday's much heralded general methods have a sad way of becoming today's awkward memories!

In fact, attempts to define classes of similar interventions by apparently well-focused descriptions like 'inventory-control' or 'distribution planning' or 'production scheduling' have shown those classes to have too great a diversity to be well-treated by similar interventions. Moreover, attempts to span those classes simply by subdividing them have again come to grief as exceptions are discovered in particular industries. So it would seem wise not to describe classes by anything more general than particular functions in particular industries, and even then to be prepared to return to the level at which we regard each intervention as having substantial new features.

That is not to say that we shall not find general theories of intervention useful, or that we shall not find general proposals for intervention worth considering. It will depend on what sort of would-be-scientific intervention we intend. It is a mistake to look for general prescriptions for classical intervention into essentially new areas: both Phillips and Feyerabend have fired broadsides on that topic. On the other hand, if we are wary about the traps we can make for ourselves in defining classes, we might offer useful prescriptions against an assumed background of intervention experience provided that we do not regard the prescriptions as a set of sufficient instructions.

Perhaps the message of this section is that, for all the support we have from societies in general and OR communities in particular, we are essentially on our own in each new intervention. It is in that unique situation that we try to understand and act in terms of what we already know and of what we can discover. There is a place for each. Even in the classical beginnings of wartime OR, the participants could wryly voice the distinction between the physicists who were inclined to advise on the basis of what they thought they knew, and the biologists who were inclined to go and see!

8.2 Clarity, precision, and changing concepts

In the first section, I emphasized the uniqueness of our individual experience. But I did so in the expectation that that uniqueness was not a complete barrier to communication between us. We do much of our communication through marks and sounds, and it is evident that they evoke corresponding mental images surprisingly effectively, even if the boundaries of what is evoked are far from sharply defined.

For much of the time it does not occur to us to question the meanings we extract from a sequence of marks or sounds. Of course, there are many times each day when we are not sure what is meant, or when we feel that what is meant is incorrect or unacceptable, but these are not occasions when we feel we have completely lost our grasp on the drift of individual words. For example, you can understand every word in this trouble except the one I mistranslated into French and back. Indeed, if we do not treat the meanings we extract as reasonably reliable then we have no language to work with. So we are inclined to treat as reliable whatever sense we extract from word sequences unless any part gives us qualms. That is, if we feel no doubt, we believe we have understood what is said. Whether that decoding produces a reliable experience of meaning corresponding to what was encoded is another matter. And even if it does, it does not necessarily give a reliable indication of what it purports to do, since language can be used to deceive as well as to inform.

The extraction of meaning from word-symbols is not a straightforward matter. On the one hand we have word-symbols and sequences of word-symbols which we can see and hear and capture for indefinitely long periods in physical form, and on the other hand we have the different collections of mental images associated with the word-symbols in the minds of different actors at different times in different contexts. Insofar as the collections of images are common we have a sort of objective world of meaning which is both distinct from the actors and distinct from the physical symbols in which the meaning is expressed. This idea of Popper, that objectivity derives from inter-subjectivity, is supportable at its most dramatic by noting that the same meaning (or approximately the same meaning!) can be represented by a wide range of languages, and can even be gleaned from the written language of dead races to which no key initially exists.

But insofar as the collections of images are not in common we have inter-subjective problems of consistent meaning whether noticed by the actors or not. Indeed, one aspect of articulate intervention can be the discovery of hitherto unsuspected differences in meaning: for example, in one factory three quite different interpretations were given to the statement of planned factory output, leading to unmatched production rates! A dictionary gives some indication of the variety of meanings we ask a word to bear, and it can be argued that the meanings that get into the dictionary represent artificially sharpened distinctions from a much more finely differentiated field. Moreover a word carries with it an extensive field of associated images, some latent in that they will come to the fore in appropriate circumstances, some hazy and indistinct and likely not to come into clear conscious focus, and all associated with clusters of feelings with which they interact. So, following Foucault's discussion of language, we can say that although we are often in no real doubt what is intended, the field of images associated with a particular word for a particular person is indefinitely large, vaguely defined, activated differently on each occasion it is evoked, and modified in its extent and connectedness on each occasion it is evoked. It is possible to change the associations of a word or a short sequence of words for other people quite deliberately: Priestley once wrote a play so that the audience would feel a chill of horror when the phrase 'Anyone for tennis?' was heard; at a more serious level, factories develop their own private versions of English; and under steady propaganda a language can lose the power to express certain kinds of concept. The unfixed nature of meaning implies that if ever we tried to say exactly and fully what we meant on a particular occasion of utterance, every word would need its own cascade of clarifying explanation, and each hearer would make something different of it.

At one end of the scale, the richness of association is related to the making and enjoyment of poetry. At the other end of the scale, the evidence from the natural sciences is that we can sharpen the definition of the associated set of images, or select sharply defined subsets, both to transmit meaning reliably and to make statements which correspond closely with reality; statements which allow the consequences of some actions to be closely predicted under controlled conditions. The ambition of the social sciences would seem to be to follow the example of natural sciences in being able to transmit fairly precise meanings which are a fairly accurate reflection of reality and which allow fairly precise predictions of the consequences of actions.

The clarification of meaning, by attempts to be clear to oneself and to others, makes use of further symbols to point out distinctions that might otherwise not be made. Once we have seen how much progress we can make in the clarification of meaning, we might feel that it would be easy to reach the limit of absolute clarity. Indeed, if we have some familiarity with the language of mathematics we might think of the process of clarification as being rather like the process of converging to a limit. That would be misleading. Symbols point to changing collections of images, even if in some cases the changes are mostly small and infrequent. Moreover, we run into problems with the indefinitely large cascades of further symbols: we cannot complete the scrutiny in a finite time, and in places we need new images and concepts to be able to proceed with that particular cascade. So exactness and precision remain ideals.

Since no act of definition can be guaranteed to give absolute clarity about the images a symbol points to, there must always be some uncertainty about mapping from symbols about the real world to symbols which are hypothesized to be manipulable as though they were absolutely precisely defined. Whilst such a mapping enables all the power of mathematics and computation to be released into the argument, the correspondence between the conclusions of an argument in precise symbols and the real world events they point to depends not on the logic of the argument but on the precision of the original symbols about the real world and the correctness of the statements that were made with them. Moreover, in the act of mapping from the original symbols to the would-be-precise symbols we may not notice the variety of meanings that have been condensed into a single abstract meaning and which may spill out when we map back from the precise conclusions to the real world. How can we be sure that all the shades of meaning that we give to 'profit' will be reliably wrapped up into a single precise symbol, reliably processed by a single precise argument, and reliably mapped back into all the corresponding cluster of conclusions that would have been drawn had we treated each shade of meaning separately? Are we altogether sure whether argument in precise symbols about economic principles is a powerful generalization or a misconceived aping of natural science? It is after all not argument about the presumed regularities of nature but about the theory- and proposal-saturated variety of human affairs. And is that variety not something that is continually giving us surprises about how our theorizing does not quite fit particular instances?

Whether we argue by precise symbols or by natural language, we may become aware of the deficiencies of the image field we currently associate with the symbol. One way in which this can arise is by trying to embed the symbol in an argument. If after two or three moves in the argument we find that, whilst there was initial agreement about the opening statement there is nevertheless strong disagreement about the validity of conclusions that have been reached, we can suspect that we did not in fact start with a symbol whose image field was well under control. I shall argue in a later chapter that some of the main problems of articulate intervention arise from the central place of vaguely defined symbols in our industrial civilization, but that vaguely defined symbols seem to be an intrinsic feature of large-scale industrial culture—that social culture seems to require them. Here I am only concerned with the improvement of argumentative reference where we intuitively expect to be fairly precise.

It seems that an awareness that a symbol has deficiencies is a useful precursor to new concepts. By concept I have in mind whatever it is which is indicated by a willingness to view part of an image field as a unity. New concepts are then whatever is indicated by a sense of unity about new image fields, new contents of an image field, or new groupings of an image field. If we accompany this newly-sensed unity by articulating symbols pointing to it, then we may make it non-fleeting and communicable. New concepts do not arise easily, but they can often be easily communicated. The ease with which they can be communicated is no reflection of the difficulty with which they were achieved. It seems that articulation can in some circumstances bypass slow mental processes that would otherwise take months or years, if they occurred at all. You may for example be aware of the principle of optimality, which is the central notion of the abstract technology known as dynamic programming. Year by year we expect whole cohorts of students to be told about it, to reflect a while, and to say that it seems rather obvious. I find that stands in sharp contrast to Bellman's account of his slow appreciation of the applicability of that basic concept. It took years.

Something similar appears to happen in would-be-scientific interventions into particular goings-on.

8.3 Originality in interventions

It is my impression that long-established in-company and in-industry OR programmes depend in part on the slow and occasional generation of new concepts about their sponsoring programmes. Over the years the generation of new concepts is a critical reason for their continuance. The concepts may not be spectacular, and may be of little interest to those outside the sponsoring programme. They arise in response to a slow realization that the sponsoring programme lacks or is mistaken about some ground concept. The realization may well occur during the course of studying related matters and striving for concept clarification. So, for example, an OR group may be responsible over the years for changing the terms in which capital investment is discussed in an industrial organization.

No programme has a monopoly of new concepts. It is misleading to represent would-be-scientific intervention as though it is the intervening programme that offers the new concepts to an otherwise passive entered programme, though it might reasonably be hoped that in some aspects of the entered programme it would be the main source. But that would perhaps not so much be because of some inherent general originality, as because the current concepts and theories in the intervening programme give a particular direction to its heuristic potential.

Since new concepts are a slow matter, an intervening programme must rely to a great extent on its stock of available concepts and those which can be introduced from the entered programme. This means that the intervention creates a new mix of concepts which, with the related new mix of theories and proposals, can lead on to new theories about the entered organization. The experience is creative and exciting and a particular form of originality. Preparation for theorizing in particular situations requires educational programmes to provide concepts and experience of articulating theories of the particular in terms of those concepts.

The process of theory-generation might well include thousands of half-formed theories with lives from a few seconds to a few days. The process of formulating and criticizing theories of a particular situation can be every bit as demanding as research on general matters, and in some ways more so since OR interventions usually include the proposal that the intervention is about advice on action, and that in turn implies a search for a wholeness of view rather than being content to make correct statements about some aspect. Moreover, the standing presumption of the natural sciences that there is a regularity to natural phenomena does not carry-over easily into studies of social phenomena. That is not to claim that the implied intellectual challenge is always met, or that it is even recognized. The magnitude of the implied challenge is daunting, the time is limited, and the emphasis is on some kind of beneficial action rather than on theories for their own sake. And that means that criticism, particularly extended experimental criticism, of theories about particular aspects of real goings-on or of features of abstract models, is limited. There is perhaps no great harm in that, provided that both we and our sponsors realize that our behaviour is not markedly different from our sponsors' and that our theories are subject to change—that we hope our recommendations for action will be satisfactory, but only time will tell, and there may need to be revision. In this matter there are large differences between OR groups: some assume a great deal of responsibility for firmly establishing new theories, whilst others seem content to offer themselves chiefly as agents for handling the theories of their sponsors.

9. Two self-concepts of OR

9.1 'Science' and 'Optimization'

Histories of the OR world see it as having continuity from the early days of the second world war. At that time two self-concepts were established. First, there was the concept of the critical investigation of actions, theories and proposals about matters of some importance to large sponsoring programmes: this was later to spread from military areas to basic industries and thence to industry and commerce in general and on to public and administrative activities. Secondly, there was the concept of the abstraction of precise-symbol problems, which under the influence of RAND and other government-funded research organizations in the USA became a separate activity which was later closely linked to the availability of computational support.

Some of the early activities of uncovering an understanding about particular classes of real situations were clearly scientific in their style in the spirit of observational sciences, and this has led to the continuing wish to be as science-like as possible and, as we saw in the first chapter, to make a charter statement. Much of what was attempted in the use of precise-symbol representation could be described as optimization: the search for some mix of actions that would maximize the value of the resulting consequences to the sponsoring programme. The free use of the words *science* and *optimization* has had an unfortunate history in the OR world. Because most of its generally available writings have been about the internal challenges of precise problems, the free use of the two key words in the literature has largely associated them with the internals of precise problems. Since there is a resemblance between the precise-problem literature of choice and the precise-problem literature of natural science (e.g. the similarly structured writings on posing a precise problem of mathematical physics and then turning one's attention almost wholly to the problem of mathematical deduction), and since much of the precise-problem theory of natural science is known to fit nature closely, there has also been an uncritical tendency to regard the models of the literature of precise choice as sound models of real decision situations, as being ready for easy use in deliberative argument, and as forming the main intellectual challenge of OR.

The evidence is different. Many of the precise problems of the literature are defective in that they correspond to no possible point in the sequence of deliberation in an action programme! The issues they embody are often a sharply delimited set, and the quantities symbolized are treated as though they were known precisely and as though that knowledge were not subject to revision. Precise models need to be separately checked in every situation they are used in, and they can be used quite misleadingly. But if their properties and shortcomings are understood they do have a powerful contribution to make to deliberative argument, if the problem of embedding their use in programmes can be solved.

So how far do we manage a science of the real world and how far do we manage real optimization? What sort of choices can we make for our activity within the OR world? The potential is wide. Part of my purpose is to provide a way of describing OR activity which will free OR communities to move into new areas without feeling too naked! But I feel that to serve that purpose I must try to exorcize some of the ghosts!

9.2 Origins in science

What was the input of science in the original OR programme in the UK? Most tangibly it was the transfer of a considerable number of scientists and mathematicians, many of them outstanding, from research programmes in natural- and life-sciences to the investigation of various wartime action programmes. A simple view of this, with some element of truth in it, was that it represented a transfer of considerable powers of originality which could be relied on to show some kind of rapid and penetrating originality in new situations. Whilst I consider that the wartime programme would not have been achievable by any other group, I think one must look more closely at what they did and did not bring to the situation, since the picture of general originality has later been rather misleading when claims have been made for the general utility of OR and other forms of would-be-scientific investigation. Two aspects need attention: the theory content of the research programmes from which they came, and the proposal content of those programmes.

The scientists as outsiders brought minds well stocked with concepts from the natural sciences, and supported by the considerable interpenetration of mathematics and natural science. The most important concepts were not so much the advanced ones from the research programmes they had left, but quite simple ones like the pace and inner momentum of natural processes, the visualization of measurable quantities as interconnected and therefore able to drive each other through mechanisms representable by algebraic relationships or by graphical presentations, the degree of smoothness that one might expect in such relationships, notions of continuity conditions of the what-goes-in-must-come-out type, and the idea that what is apparently haphazard can sometimes be the result of quite stable probabilistic mechanisms. These quite basic concepts, models if you like, that surface events can be thought of as driven by measurable and describable processes and interdependence relationships lying not far below the surface, were carried into a variety of novel situations with great effect. They were welcome where there was a gap in current conceptualization, but had a more mixed reception where there was already strong conceptualization not yet considered to be in doubt in the military programme. The search for measurable variables and relations between them was apparent in the problems they perceived they could tackle and in their effectiveness in doing so.

The proposal content of their previous research programmes had its effect too. Again, the most important proposals were not those about the particular kinds of advanced exploration and experimentation guiding current research in the programmes they had left, but quite simple ones like the proposals that understanding would be sought in the heuristic activities of observation and recording and analysing data, that theories should be about observable and measurable consequences, and that retrodictions and predictions should be checked against observables and measurables in ways that would show any systematic error. This was unlike everyday proposals for the generation and criticism of theories: there the proposals for quest and criticism are many and varied and inconsistent and in many cases the proposals are bound to fail or be inconclusive because of the subject matter attempted.

By contrast, the standards of criticism of theories in scientific research programmes were fairly well defined and fairly coherent, and embodied proposals that there should be a logically consistent deductive argument from theory to observations: an argument which should be free of evident error in its supporting theories. From Waddington's account of studies in aircraft readiness we can see that this meant a readiness to embark on some most complicated structures of argument to check that the great variety of observables did not indicate some internal inconsistency, and an unwillingness to shrug off what would not fit. Again, this affected the problems they perceived and their effectiveness in tackling them. It was particularly appropriate for classes of repeating situation where theories built up in response to earlier examples of the situation could be checked-out against later examples, monitored while proposals for improved action were in use, and used to detect their own dwindling validity as the situations changed. Such situations provided opportunities for classic examples of the full range of the methodology of natural science in action.

9.3 Unnatural science

The proposal content for the active scrutiny of theories against the background of actively sought observables is a high level of achievement for a species. It contrasts with the everyday handling of inconsistencies and the search for them. Festinger produced some concepts for theorizing about what happens when an individual has inconsistent cognitions, or perhaps not so much when he has inconsistent cognitions in the view of someone else as when he *experiences* his cognitions as dissonant. He may like it and look for more (some people seem to enjoy living with conflicting perceptions), he may act so that the externally induced cognitions are changed to reduce dissonance (e.g. by physically leaving a situation he does not like), or he may reduce dissonance by generating for himself within his own mind new cognitions which, by replacing or supplementing earlier cognitions, lead to a sense that the dissonance has been reduced (e.g. that next observations are going to be what he thinks, even though the present observation is not what he expected). Our view of his new cognitions will depend on our own view of what is acceptable or desirable. We might variously think of the arrival of the new cognitions as wishful thinking, realism, showing adaptability, rationalization, showing creative insight, devilish cunning, stupidity, or as being heavily conditioned by the language and traditions of some sub-culture.

We often experience cognitive dissonance when we have no immediate explanation for some event: the it-must-be explanation that we offer ourselves serves to reduce dissonace, but that does not normally lead on to critical scrutiny unless we enjoy the challenge of criticizing our own views or unless we are with others who enjoy criticizing ideas. So it seems that in everyday affairs we are inclined to accept a reduction in cognitive dissonance from any convenient source. By contrast the research programmes of natural science have proposals that we should pay particular attention to the logical dissonance between theories and to the observable dissonance between theories and observables, and that we should actively seek to sharpen our awareness of such dissonance.

In postwar years, with the return of many of the most able scientists from the general OR programme to their original research programmes, the science input to the OR programme changed. There was still an inflow of people familiar with the concepts of the natural sciences, but more through learning of them within educational programmes than through using them in any action or research programme. And since educational programmes concentrate on the theory content of research programmes to the virtual exclusion of the proposal content, there was a general loss of emphasis on the pursuit of a critical methodology of theory appraisal. Indeed, the postwar situation was probably worse than would have been expected simply from the declining first-hand experience of research proposals since, as Lakatos has argued, there is a history of presenting them in educational programmes in severely garbled form, in part because the proposals of research programmes, whilst implicit in their development, have not been well described in the research programmes either! It would be wrong to present the loss of critical methodology as sudden, since up to the period of the general availability of computers it was prized wherever it was evidenced, as can be seen from the awards of the English-speaking national OR societies, and from published case histories like that of Edie or those in the books edited by McCloskey and Flagle. But the loss of critical methodology was enough for the Operations Research Society of America to broaden its criteria for the Lanchester prize by 1959.

More importantly, however, whether or not the personnel of the general OR programme had changed, the critical methodology of the natural sciences is quite simply inadequate for many of the interventions that have been undertaken. For example, interventions which address the problem of acting on theories whose correctness has no possibility of being checked until several years after the actions have been taken, or interventions into situations where there are conflicting proposals on the evaluation of consequences and no simple loyalties to provide an easy means of resolving the choice. So what I think we have in part been witnessing is the slow business of getting to grips with the problems of devising patterns of criticism, of constructing critical methodologies, for those areas not readily dealt with by the methods built up over so long a period in the natural sciences. At the same time we have been witnessing a stream of interventions, with such methods as we already have, into many different action programmes and into many different classes of action programme. It is the intervention into classes of action programme which most clearly reflects the science-like concept of OR, but which at the same time shows the concept to be in need of extension.

9.4 OR as mathematics

Within the natural sciences, the precise symbolism of mathematical forms and arguments has an important place in the expression of theories and in deducing the consequences of theories. This enables sharp precise predictions to be made and checked as part of the science programme. It appears also to allow sharp technological predictions to be made, but since technological actions lead to whole physical realizations it is not uncommon to find that, despite the precision, some aspect of the whole has been completely overlooked. So suspension bridges have been known to twist to pieces in the wind, and nuclear reactor elements to fail in three years rather than thirty, and metals to become embrittled in unexpected ways.

In addition to providing an expression of theories, the precise symbolism of mathematics is also used in OR with the intention of supporting the choices to be made in deliberative argument. The hope is that much of the complexity of deliberative argument can be largely or entirely replaced by computations which are a direct result of precisely expressed understanding. Indeed, since the OR world has produced few, if any, mathematically expressed theories about identifiable *general* classes of real situation, and since the discussion of the construction and criticism of models in particular real cases is weak, the place of mathematics in OR programmes is almost entirely concerned with:

(1) The deduction of consequences in models for specific individual interventions,
(2) The speculative deduction of consequences in models guessed to be of potential use in individual situations,
(3) The abstract technology of choice in specific individual interventions,
(4) The speculative provision of abstract methods of choice guessed to be of potential use in individual interventions.

So almost all the mathematical content of OR can be conceived as lying within the framework of reflection before action. That is a framework which also includes the processes of conceiving possible actions, the processes of conceiving theories about particular situations and the consequences of actions in them, the checking of theories against reality as far as that is possible, the modification of theory, and the choice of actions to take. We can place the role of mathematics within that context: the deduction of consequences, the choice of action, and the adaptation of actions within undisturbed sets of theories and proposals are evident areas for the use of extended precise formal logic, if it is to be used at all. Reflection before action may not however express all that is necessary: it may not remind us adequately of the many interpenetrating action programmes and their conflicting and logically irreconcilable proposals, nor does it suggest that it is the processes of reflection which are under study, nor does it adequately reflect the speed and vigour of deliberative argument, in which peripheral theories and proposals in participating programmes are subject to rapid review and change. Nevetheless, the concept of articulate reflection before action is a concept which embraces nearly all the precise-problem content of the OR world.

There is no doubt that precise-symbol problems, abstract problems, have been an area of great interest and usefulness in the practice of OR, and have been an important influence on the self-concept of those who see themselves as members of the OR communities. Indeed it is rare to find participants who have not had a considerable diet of mathematics in undergraduate programmes in their original subject or, then or later, in programmes of OR education. But I am concerned here not with the internal puzzles of particular classes of precise-symbol problems, but rather with the structure of their relationship with articulate reflection in action programmes.

9.5 Mistakes and opportunities

All kinds of precise problem share whatever virtues and vices stem from the implicit property that the problem begins at the point of statement. The mapping of real action problems on to a precise problem is an action external to the precise problem and is in no way included in the statement of the precise problem. That would be a facile remark were it not the case that the distinction between real problems and precise problems is often overlooked, particularly in educational programmes and in programmes of selling what are claimed to be ready-made computer methods. The problem is serious because the word-symbols used to conjure up statements of real problems are also used to conjure up statements of precise problems, and it is not common to label the symbols to remind the discussants whether they are dealing with conjectural statements about the real problem or with logical statements about the abstract precise problem. Since these two are often not distinguished, there can be some curious mistakes.

The risk need not be great for the person working on a precise problem which has been abstracted from some goings-on of which he is a part, since whatever he concludes has at least some chance of being compared with the reality and being criticized by others who know that same reality from their own very different points of view. The risk is much greater for the person working on a precise problem which has been guessed to be of potential use somewhere. For him there is neither the reality against which he can check the components and conclusions of his theorizing, nor are there the participants in deliberative argument through whom he can appreciate the possible modes of relationship there might be between his abstract model and ongoing deliberations. If he has had plenty of experience of the deliberative arguments in which he is intending to intervene, his imagination and memory will stand him in good stead. Without that experience there is no memory to imagine from, and perhaps no inkling that any such imagination is needed!

One consequence of that is the mistaken belief that by theory one can prove something about the real world with the same sort of finality as one can reach the conclusion of a precise-symbol argument where the symbols are hypothesized to be completely precise and in need of no further examination. So you can, for example, realize that the analyst who is talking to you or whose work you are reading is under the misconception that the desirable extremal properties of a precise action in a precise problem means that the corresponding real action has a corresponding extreme of desirable properties in the corresponding real problem, but that he has checked correspondence only by abstract argument! In principle, the claim that precise-problem conclusions translate precisely into correct real-problem conclusions is wrong. The mapping of real problems on to precise problems is always accompanied by a considerable simplification in choosing what to map: the properties of any real system are indefinitely many. So we must maintain a clear distinction between the *theory* that an extremal precise action indicates that the corresponding real action is extremal, and the *proposal* that the entered programme undertake the real action corresponding to the precise action: the theory is not justifiable, but the proposal can like all other proposals be accepted or rejected. That is by no means a pedantic comment. Would that it were! The mapping of real problems on to precise problems is far from satisfactory, and the widespread provision of computing facilities has had so hypnotic an effect on the participants of intervening and entered programmes alike that some very unsuitable precise problems have been used to choose actions.

The field of precise problems of choice has been well explored by White, whose work prompts me to make the remarks of the rest of this paragraph. One can distinguish several variations of precise problems: deducing the implications of a precise-symbol theory, deducing the precise consequences of precise actions, deducing the precise actions to give particular precise consequences, and deducing precise actions to give extremal precise properties or near-extremal precise properties. Some of the problems are decidable problems in that the problem statements have in principle a single answer entailed by the problem statement. For example, what is the probability of shortage given this precise problem, or what is the maximum profit given that precise problem: 'probability' and 'profit' are precise entities following from the premises of the precise problem, and no claim for their relevance to the real world is, or could be, made *within* the precise problem. The field of decidable problems can be augmented by introducing partly decidable problems, which impose precise constraints on the possible answers without entailing one in particular. It seems likely that such augmentation will find an extensive place in the general OR programme, since it permits the combination of precision over some areas of a problem with choice and exploration in others: Tocher has remarked that exploration would be driven by the device of introducing decidability by some further conveniently controlled constraint.

9.6 The risk of programme degeneration

The wide range of internally interesting puzzles in the solving of precise problems has both advantages and drawbacks. The internal puzzles have proved to be an open-ended field of great complexity which has generated its own research programmes of the theoretical solution of those puzzles, research programmes on computational methods for solving the puzzles, and action programmes for providing computer realizations for solving the puzzles. It is possible for participants in those research and realization programmes to be quite unaware of the way in which precise problems can be embedded in a process of articulate reflection. In the UK this was a specific concern of McLone's report to one of the Government's research councils, though he used his own language rather than the terms I have just used.

Some research programmes are now proceeding by the generation of their own precise problems in a steady stream without reference to particular real situations. In the OR world and in cognate areas this seems to be true of various research programmes in mathematical programming, mathematical economics, probabilistic processes, and control theory. Although this makes for ready advances in generalizing formulations and in developing increasingly abstract levels of concepts in which to embed the discussion of lower-level classes of precise problem, there is a major disadvantage. The potential experience of dissonance between theory and reality can be avoided, since a group concerned with self-generated precise problems can also generate for itself such theories and proposals as ensure that its attention is focused on the more limited logical dissonances that occur within the course of its activity; and, within that set of logical dissonances, that its attention is focused mainly on those dissonances which there seems to be a good chance of overcoming with the present core of the programme and its auxiliary theories and proposals.

Now there is no reason why a research programme should not be developed in any way which is consistent with its own proposals, insofar as it is not actually prevented from being so by the actions of others. However, it is misleading to confuse the *proposal* for independent development with the *theory* that independent development is the best way of achieving greatest usefulness to the OR world, or with the *theory* that independent development is the best way of achieving greatest usefulness to the world at large. Because the participants of self-sustaining precise-problem research programmes risk missing the dissonance between reality, real theories and precise theories, there is a risk that the research programmes omit the trickiest problems of action and there is a risk that the research programmes degenerate away from the precise problems originally tackled to precise problems from which the awkward features have been removed and which are therefore further away from modelling any real problem. It is well worth reading Lakatos on this.

The risks of programme degeneration have little to do with the sophistication of the mathematical or computational methods used. The risk of programme degeneration does not run from a 'high' for the most abstract mathematics to a 'low' for straight arithmetical methods. Nor do the risks run the other way! The risk is related to the core of the research programme: on the extent to which it is no longer concerned in part with mappings from real problems, and on the extent to which action proposals for real problems are considered. In any precise argument, the collection of precise premises one starts with is central: if these fail structurally to represent any real situation in deliberative argument, then the rest of the argument is unlikely to be usable whether it is simple or sophisticated. I am as unimpressed by some apparently earthy approaches to production planning as by some of the more ethereal probability models which cannot possibly be tied down with the data available. But equally, what I value ranges from the simple to the advanced. The cut is not to be made by the complexity and abstractness of the analytical methods in the precise problem but whether the precise problem models anything well.

9.7 Some trial statements

Since the concept of OR as an activity of precise-problem solving is defective, we might try to put something more complete in its place. I am not inclined to believe that anything worthwhile can be wrapped up in a definition which will serve for all readers and listeners on all occasions, but we might try this as a view:

> OR is the practice of intervening in action programmes to provide articulate reflection on action proposals, looking in particular for the possibility of taking parts of real problems and treating them as precise problems within the sequences of reflection, either to resolve them as they arise or, if there are problems which repeat with similar structure, to devise methods of conveniently resolving the precise problems that are expected to arise on future occasions.

Such a concept of OR might reflect quite well the expectations of the participants in intervening programmes and entered programmes. Certainly I meet many young OR people who find something along those lines a satisfactory clarification of their role, who expect education in OR to be oriented to such a role, and who would for the time being ask no more than to enjoy their role on those terms.

But insofar as that description puts little emphasis on the active search for reliable understanding of reality, it hints at abdicating responsibility for the science of the situation. And it is that hint of abdication which I find most worrying when I read accounts of modelling which do no more than reflect the current views of the participants in the entered programme. I am not alone in that. Some of the OR programmes and sponsoring programmes of longest standing would be seriously limited if the precise-problem orientation of OR had been so emphasized in their development. For the participants of those programmes, a wide non-constricting view would be needed, perhaps along the lines of:

> OR is the practice of intervening in action programmes to address any significant problem of articulate reflection

> before action, and by any conceivable method examining and adding to and modifying the theories, proposals and repertoire of actions for the entered programmes and other programmes with which it interacts, so as to permit a clearer understanding of reality, and a more reliable and timely choice of action.

The intention of a definition of that breadth is that articulate intervention be seen as concerned with the full range of concerns of the participants in an entered programme, insofar as we can usefully articulate views about them. But it is a definition which would equally well include several other groups of highly able people who are concerned to make their own distinctive contributions. In effect I am suggesting we do away with demarcation disputes and that, if we do want to draw tighter boundaries round ourselves for the purpose of self-definition, we do so voluntarily and temporarily rather than with the idea that there is something permanently correct about the present delineation of our field of activities.

My querying of the limiting of OR to precise problems had, however, a serious theoretical purpose within the development of my argument. Action programmes usually contain core proposals that defy precise statement. I therefore think it may in principle be unwise to regard the development of OR as being characterized by combining theories of the consequences of action with proposals for choosing actions. Indeed, in a small sample of companies in the Midlands in 1976 more than half the OR activity was about providing models for the display of consequences but not for automatic searching over a range of actions by optimum seeking methods: that is, there was an emphasis on What-if? questions rather than Which? questions. So, rather than conducting articulate intervention under the implicit assumption that what will emerge will be semi-permanent methods of taking over sections of deliberation, it might be more fruitful to say something like:

> The thrust of OR is towards the open and debateable display of theories, proposals, repertoires of actions, and estimates of consequences. The intention is that the thrust be supported by ingeniously and imaginatively

contrived enquiry into action programmes, abstract systems, and the engineering of heuristically modifying interactions between them.

I include the open and debateable display of proposals with some trepidation, since it seems to me to represent a potentially much more radical view than simply the open and debateable display of theories. I take that view because I think that the major action programmes of our industrial civilization contain core statements of considerable vagueness but which play key roles in the achievement of social cohesion and in the attainment of large-scale social and technological innovation. I am therefore unclear about the consequences of a sudden major increase in clarity: it could conceivably weaken the impact of passion, and leave a society vulnerable to invasion and oppression. Perhaps that is yet another example of there being no guarantee of benefit in any individual instance. On the other hand, what sort of clarity would it be that overlooked the savage hostility that can erupt at all levels of civilization? What sort of science of decision would it be that overlooked the combined use of force and advocacy as the framework of national and international affairs? So, onward to clarity!

10. Conceptual problems from society

10.1 A muddle of meaning

To reflect articulately on the present practice of would-be-scientific intervention and about its future development is much more difficult than might at first appear. The conceptual environment of would-be-scientific programmes, like the conceptual environment of all programmes, is messy, since the general standard of conceptualization in our civilization is poor. By that I do not mean that I think that people in general are dim, since I meet very few people who do not have shrewd things to say about their own affairs. Rather, I think that we are all in the middle of a muddling experience of word-symbols insofar as they refer away from our immediate familiar activities. The obscuring and frustrating experience of word-symbols is one from which we can only with difficulty extract ourselves, even if we agree that it is in our interest to do so.

In part the muddle is historical in origin. Our societies have changed enormously over the past several centuries. The *Weltanschauung* of national groups shifts considerably from generation to generation, and is greatly differentiated within national groups. The changes have been accentuated by military and mercantile prowess and distress. The preoccupations and emotional responses of whole populations can be systematically shifted in major ways by the mass media and by specific or general violence, whether the shift be intended or an unintended byproduct. Moreover, there is variation between individuals, and within one individual over time.

The result is that each of us has a large collection of unexamined concepts and articulations of uncertain pedigree picked up over a lifetime and strongly reflecting the social conditions in which we have lived. We additionally have more orderly collections of concepts from areas in which we have received some degree of coherent instruction. And finally we have a few concepts we have examined with thoroughness for ourselves. The concepts, and the symbols we use to indicate them, are associated with patterns of deliberation which themselves range from the unconsidered to the highly scrutinized. In a professional role we might prefer to adhere to the well considered, but in our everyday roles we use all of them.

So in deliberative argument, the whole mix of concepts is around. To an outside observer, the mix of concepts in an entered programme may appear to be riddled with inconsistencies or ambiguities. But for each participant, his own collection of concepts may well exist without giving any notable experience of dissonance. The whole collection can potentially come into play during articulate intervention, and that will include any notions of assessing desirability.

10.2 Motivating symbols

When desirability is examined in the normal course of social interaction, the word-symbols are usually used in a way that suggests that at least within the reference group of the speaker there is the possibility of an agreed view of what is desirable and what is not. If we look further at the debate we are likely to find a tension between symbols pointing to personal needs and symbols pointing to the needs of others. On the one hand there are considerations of food, shelter, security, pleasure, property, and self-fulfilment; on the other hand there are cultural concepts like fairness, justice, duty, opportunity, service, and freedom.

There are also in our industrial civilization many symbols which point to particular action programmes which are allowed great claim on the individual: state, family, money, the party, profit, the union, the company. These symbols, and others like them, have some curious properties: in addition to their straightforward utility for indicating some broad classes of social programmes, they are powerful symbols for cohesion, and powerful symbols for conflict. They are virtually unusable as terms in a logical argument, and they almost totally obscure any clear consideration of the direct and indirect effects of actor upon actor. Naturally, these motivating symbols have not reached their present powerful place without widespread and repeated use in many action programmes. Their use is associated with strong and even passionate feelings about the desirability of particular classes of action, theory, and proposal, and their use is often in the context of force or advocacy.

The cohesive power of the great motivating symbols is quite remarkable. By them we can be persuaded that there are extensive areas of common interest and we are persuaded to act in ways which are fairly predictable and which are consistent with the extremely complex interwoven arrangements of our civilization. So to look at society as though it did no more than reflect agreements freely entered into is to use a deficient model. It is a model that might possibly serve to describe the setting up of negotiated arrangements among a newly met group on a desert island, but it ignores the dominant place of existing language in a continuing culture.

When desirability is discussed in terms of general motivating symbols, with all their power for cohesion and division, it seems a little other-worldly to suppose we shall be able to organize our affairs by discovering that individuals have some stable intermediate measurable utilities, and to suppose that we could somehow be saved from the messy juxtaposition of our perceptions of possible consequences, our unfinished debates on desirability, and our imagining of new proposals.

10.3 Pluralities

It is perhaps an unconscious choice that has set the OR world and other groups of would-be-scientific interveners mainly to consider matters which take the basic coherence of society for granted: to assume that it is appropriate to concentrate on whatever present conflicts are visible, and not to consider whether there are problems of maintaining such areas of agreement as already exist. So, for example, proposals for studying men at work to improve their efficiency can bring work to a halt, a result counter to the expressed aim. In this matter I think that sponsoring organizations are sometimes wiser than interveners in being suspicious of the assertion that the evident purpose of the sponsoring organization is to make some readily recordable measure of its activity take on as extreme a value as possible.

On the other hand, our society now has many institutionalized programmes which do contain proposals that the effectiveness of their sequences of actions be indexed by some simple measure. So the symbols pointing to the simple measures and the symbols pointing to the institutionalized programmes have become powerful motivating symbols in their own right. For example it is regarded as enough that, because the 'costs' of a 'department' can be cut, it is alright to cause a long period of distress and disorientation to a person. Despite the savagery which the motivating symbols are understood to sanction, the symbols are nevertheless held in high regard by many more than those whose interests are immediately served by them. They are a necessary part of adjusting our varied programmes to a changing world, and counter-proposals for permanent participation in a programme are hardly well-adapted to change.

Much of the success of the general OR programme has been in intervening in action programmes where there were standing proposals that some simple measure should index achievement and choice. The entry into pluralistic situations implies a major change. There the main pluralities are all to be held in regard, and it is unlikely that there can be any potential proposals where all the pluralities can gain in terms of their starting perceptions. For the pluralities themselves, progress entails modifying in at least some of the pluralities the set of motivating symbols and their effects. This points to OR as an intervention to produce mutual clarity, rather than to provide some superordinate source of symbols: as an intervention to produce improvements in the means of debate and the resolving of conflict, rather than to produce assertions that all the pluralities will turn to in preference to their own!

10.4 Handling disagreement

The divisive power of motivating symbols, whether they point to institutions or to measures of achievement, rests on their repeated use in many action programmes. But it is repeated use which in different social groups is characterized by significant non-overlap with the image-fields and emotion-fields the same symbols are associated with in other social groups. This makes the symbols of little use for debates between the groups, but great for fights. And for both groups the associated image-fields are likely to be large and ill-defined. Although clarity of reflection *can* lead to passionate commitment to action, it is more usual to find passionate commitment to action engendered by, or at least supported by, the use of unclear motivating symbols.

The lack of clarity seems to be a consequence of the enormous image-fields that are pointed to by symbols for institutions and simple measures of them. For example, the word 'profit' points to such an enormous range of situations with such an enormous range of characteristics that it is virtually useless for conducting debates. What may be assented to with one part of the image-field in mind may prove not to lead to similar assent when another part of the image-field is activated. Since large-image-field symbols are often used for conducting debates and for declaring proposals intended for widespread application, we ought to retain a clear view that there is no logical justification for claiming that all possibilities have been adequately covered in deliberative argument. It is a common experience to find that counter instances are only recognized when they are later experienced by being met with for the first time by the deliberators or by others they wish to be bound by their proposals.

Since formal agreements in the forms of laws, regulations and contracts form an essential part of the proposal structure of industrialized societies, it is perhaps understandable that they are presumed to have been announced, understood, and accepted in terms whose extent and content are in principle the same for each actor. It is a truly remarkable presumption, and it is perhaps surprising that so modest a part of the legal programme is taken up by resolving disagreement.

A much more basic form of disagreement, however, is not over the interpretation of current proposals, but whether they are to continue in force and what new proposals are to be agreed sufficiently to make them implementable. In conducting debates on general proposals, the participants are concerned perhaps more to manipulate utterance to get the effect they want than to participate in anything that presumes a complete common interest. Would-be-scientific advice to a politician would then need to deal not so much with the content of what was said as its likely impact on various classes of hearer: not so much the relation of meaning to utterance as the relation of feeling and action to utterance. So we get sonorous strings of word-symbols which use the language of theory and proposal, but which are designed for their affective impact and which say little: that is the theories leave almost all statable theories still available and the proposals leave almost all statable proposals still available. So, for example, there may be a statement of government intention to 'stimulate the economy'. That says nothing: no action is ruled out, so it is devoid of content.

Specific proposals are another matter. If they have content, then people can see how they will be affected. The stimulation of the economy then might take place by increasing taxes and spending them on civil engineering construction or by reducing taxes and leaving the spending decisions to the individual. Either way, the proposals are about changing the exchange-of-obligation programme in some specific way, against the background of some attempt to engineer feelings about it.

10.5 Personifying the abstract

The engineering of feelings is partly achieved by yet another property of some large-image-field symbols: that some of them acquire image groups taken from the image-fields proper to the attributes of individual persons but not proper to the attributes of a construct. So what is indicated by a large-image-field symbol may be spoken of as though it had personal qualities. Worse still, what is indicated by a large-image-field symbol may be spoken of as though it were in some way superordinate to individuals.

Indeed we have gone even further and persuaded ourselves by legal enactment that such things as states, companies, cities and universities have some kind of independent existence: that they have needs of their own beyond simply being clever instrumental constructs for the organizing of human affairs. These are the constructs for which I most felt the need of the concept of action programme, since their development can be charted in documentation and physical consequences independently of the participants. But insofar as they are personified in their attributes, at best confusion is introduced and at worst they become sanctions for evil.

Indeed, there seems to be a widespread view that in some ways the needs of these constructs are more deserving than the needs of individual people. So we hear phrases like 'for the good of the company' and 'in the national interest' and 'saving the pound' used in ways which obscure the full range of implicit proposals which lie behind such short-hand references. We talk as though 'states' and 'multi-national companies' were in conflict, rather than talking about the particular people who are in direct or indirect conflict and who are using these extensive programmes as a way of engaging in various forms of aggression. Such use of language is entrenched, seductive, and an important factor in gaining and keeping power.

Insofar as we live in a society in which we sense there is a fair degree of justice, we would not be far wrong as seeing it as having been achieved in part by the watchfulness with which personified constructs have been treated. There are terrible examples in this century of the way in which personified constructs have been invoked to release almost unlimited viciousness and cruelty both within societies and between societies. And yet, are we ready for a society in which we can make proposals without wrapping them up in theories and intermediate constructs? Or is it that society depends on obscurity? There, I've done it myself: 'society depends' indeed! What I should be saying is something like: Some people articulate and debate their intentions, others do not try to put much into words, others keep their intentions to themselves, and all of them are likely to use personified constructs, so what would happen if they habitually aimed to think through the constructs to the individual persons beyond? I do not know. A hopeful view might be that there would be greater mutual regard. A gloomy view might be that implicit disregard would be made explicit.

10.6 Some would-be-scientific responses to societies

On the whole, the OR communities have maintained a low profile in this area, having been content to take the motivating symbols, the institution-indicating symbols, and the measure-of-achievement symbols much as they are, accepting the action programmes of society in their own terms, selecting which to support, and aiming to advise within the framework of existing core proposals in the expectation that those will change only slowly over time. There has been some attempt to transfer measures-of-achievement into new arenas, and occasionally there have been novel concepts of measures.

The OR communities have perhaps been hampered by the very things that led to success in other areas: in particular, the view of driving mechanisms just below the surface and the concentration on models of choice.

The idea of mechanism leads naturally to the view of societies as exhibiting the properties of giant mechanisms. I have already indicated, in introducing the idea of action programmes, that the giant-mechanism view has serious drawbacks. In addition it has important diversionary properties:

(1) There is a tendency for thinking to be pushed towards simply predicting the course of the giant mechanism,
(2) There is a consequent tendency for thinking about control to be limited to those actions permitted within current standing proposals on the repertoire of actions available,
(3) There is a lack of primary concentration on proposals for control by intervention, negotiation, legislation, and programmes of concept modification,

(4) There is the foolish assertion that it would be wrong to interfere with what is supposed to be a natural mechanism, as though societies were not already guided by a massive interlocking network of specific and general proposals.

In short, attention is taken away from the need to understand and influence theory-impregnated and proposal-impregnated action programmes.

It seems that there has been some history of developing choice models at the expense of developing consequence-display models, though Friend and Jessop in their studies of public decision-making, and Howard in his work on analysing international conflict, made notable progress in providing methods for the controlled steering and recording of deliberative argument about consequences and how they might be regarded by different groups. For the most part, consideration of choice has been dogged by the concept of actors completely sharing a common evaluation of consequences or actors in complete conflict: under complete conflict I am inclined to include all cooperative-game ideas since the actors seem to be in a state of temporarily suspended conflict. What seems to have been missing is a strong sense of the importance of language in human affairs: of the strong direction given by the habitual use of most concepts, against which even bold theorizing and proposal-making seem a relatively limited departure. From that point of view we can rely on some form of cooperative society if the language of its participants provides a regular rehearsal of cooperative ideas, and we can expect conflict of varying kinds according to the extent that there is a regular rehearsal of conflicting theories and proposals. Our present societies suggest that our species has selected itself by some balance of conflict and cooperation, and that individuals achieve their purposes by some similar mix of conflict and cooperation, insofar as they do. So it seems appropriate to regard human affairs neither as cooperative nor as temporarily suspended conflict.

11. Proposals

This is not a chapter of my proposals to you but a chapter about the notion of 'proposal' and its relevance to ethics.

11.1 The articulation and change of tradition

Much of a society's views on desirability is not so much indicated by concepts pointed to by word-symbols as by being embodied in the fabric of its total behaviour. What is embodied might at some point in the past have been consciously considered and then had its origins forgotten, or it might have arisen simply by the imitation of unconsidered variations in behaviour. So we have tradition in the wide sense that Popper used the term.

Within tradition we have the more restricted concept of the mores of society: those customs and conventions which are accepted without question and which embody the fundamental moral views of a society or of some of the groups within a society. Evidently there is some problem for members of a society who wish to examine the mores of their own society, since they need to establish some platform of understanding from which they can perceive and reflect on those mores. That is, they need to establish some platform from which they can search for what has hitherto been implicit, and which they have themselves accepted by following customs and conventions.

One technique for gaining such a platform is to be exposed to one or more other societies. In a more limited way there are customs and conventions within industrial and commercial programmes and these are sometimes perceptible to the outsider more readily than to the insider. Hence the value of intervention.

Tradition changes over time, both unreflectively and by the adoption of proposals. If we wish to reflect on changes that would suit us, that is if we wish to steer the development of tradition in some way, we shall need to articulate the theories and proposals which are implicit in the tradition. By that act of articulation, whatever aspects of tradition we examine are then put on the same footing as the great range of conscious and quite explicit proposals that characterize the structuring of modern societies. The proposals I am aware of include all specific requests, all stated customs within the action programmes I know, all standing conditional proposals including legislation and regulation, and all my personal resolves.

In formal terms, the opportunity which confronts an intervening programme is the opportunity to offer its theories and proposals to the entered programme. If there is to be a formal contract outlining the scope of the intervention it will need to convey:

(i) The extent to which clarity will be sought about the repertoire of actions, the theories, the proposals, and the delineation of consequences in the entered programme,
(ii) The extent to which similar clarity will be sought about other programmes which currently or in the future impinge on it,
(iii) The extent to which the intervention is expected to be a source of newly conceived actions, new theories, and new proposals,
(iv) The extent to which the intervening programme is expected to be an agent for introducing actions, theories, and proposals from elsewhere.

It would be possible for various groups of would-be-scientific interveners to define themselves not so much by announcing theories about themselves as by announcing proposals about the scope of their interventions. They might then place restrictions on the kinds of sponsoring programme to accept, the types of action, theory, proposal, and consequence that will be considered, and the range of structure and content of what might be introduced or imagined. So for example, the members of an OR community might define what they are prepared to do by listing what they think they are good at. Alternatively, they might define what they are prepared to do by listing what they think are the significant unsolved problems of their society about which it is conceivable that something might be done, and which they think they have as good a chance of penetrating as anyone else, and which they think might not otherwise be critically addressed.

So the notion of proposal has in this first section been used to suggest that would-be-scientific interveners proceed with a self-reflective sense of decision: with an articulation of the proposals they are making to themselves.

11.2 Some approaches to ethics

Traditions among other things embody notions of morality, and one kind of hope in the OR world has perhaps been that OR enquiry might in some way lead to right actions, in the sense that they would be for the good of all. That might in turn be taken to imply that there was some interest in articulating some approximation to an ideal of desirability: that perhaps investigation would reveal what *ought* to be done.

The area of ethics is in part about examining the grounds on which it is justifiable to make ought-to statements. The grounds offered as justification are varied, and include notions like some kinds of action being for the long-term good of the species and some kinds of action being evidently bad in themselves.

Even if it could be shown that something was for the general good, it would not of itself necessarily give an individual any logical sense of claim upon him. He will behave as he likes within the constraints imposed upon him, and it is not clear that he can be persuaded that he should act for the general good against his own likes. So instead of supposing that we can rely on a sense of absolute claim that an individual will respond to, most of us would be advised to take up Popper's idea and set about seeing that the constraints on others are strong enough and comprehensive enough to block the attainment of too much power and to prevent the abuse of power. Since the abuse of power against a majority often relies on the continuation of many traditional responses, it might be to the potential benefit of most of us if we replaced the idea that we are dealing with some sort of externally given ethics and legality by the idea that we are dealing with extensive networks of internally suggested standing proposals. We can then more clearly understand their instrumental nature for familiar patterns of social organizations, and more clearly perceive the shortcomings of whole sections of the network in unexpected futures.

Even if it could be shown that some action was bad in itself, we cannot rely on the idea that it would be regarded as undo-able by everyone. There is evidence that everything that is imagined, however appalling, is likely to be done by someone somewhere. Again the operational conclusion is to construct constraints, and not to rely on ethical claims having stronger impact than proposals.

And finally, even though there have been some deeply moving and profoundly influential scenarios for ideal other-regarding relationships, we cannot rely on particular theories of ideal behaviour so catching the imagination of the species that there will be a general adoption of the theories and their related action proposals. Some standing proposals are going to be necessary for those who do not adopt them!

11.3 Ethical proposals

We are in a human situation where most of our affairs are conducted by and in the light of proposals, many of them quite humdrum. There is for everyone some degree of choice, however limited. I do not say that lightly, because I am haunted by some of the extreme possibilities of that apparently mild statement: for example, the choice of killing under the palpable threat of death to oneself.

It appears that the individual's experience of the great ethical systems is that they present proposals to him, a term which I take to include commands, since commands can be disobeyed. Moreover, the proposals of the great ethical systems are not necessarily distinguished from other kinds of proposal either in the mode of presentation or in the treatment they are given in action programmes impinging on the individual. So the emotions that are associated with ethical proposals are subject to the same kinds of influence and the same kinds of modification through time as the emotions associated with other kinds of proposal.

It is true that ethical proposals are often singled out for special treatment, that they may be presented with a variety of theories about the beneficial consequences to the individual, and that they may be taken up with a strong sense of commitment to them and conviction about them. But the theories supporting them only avoid refutation by being supported by special arguments for special cases, or by extensions of the notion of benefit so that it becomes untestable within any action programme accessible to us. Or perhaps there are clear emotional benefits to be had for the individual if the proposals are taken up with enthusiasm!

Moreover, sets of ethical proposals can become an end in themselves and can divert attention away from a clear consideration of the effect of actor on actor. That is, the proposals can in time be taken to read as answers to the question: Who does what?, rather than to the question: Who does what to whom? So in the name of great ethical ideals people sometimes behave abominably.

I therefore prefer a direct line into ethics: that it is about the quality of present and intended relations between actors. Rather than starting from the notion of the needs of the isolated individual and viewing ethics as a collection of constraints on him. I think we may do better to start from the notion that social life is interdependent and look at what we might say about the basic unit of a temporarily interacting pair. It is from that point of view that I wish to reject notions of values and utilities as primary descriptors, and to introduce the notion of the awareness of proposals as a basic concept.

The great ethical systems are variations on a proposal that might well be stated in some such form as:

> You should pay concerned and imaginative attention to the other, so that by paying attention to all that he does, and by imagining yourself in his place, you can cooperate with him to your mutual benefit.

Under that proposal we can collect together much of what is found offensive under two indictments:

Lack of concern,
Lack of imagination.

To set an idea like that at the centre of one's ethical proposals is to ensure that the proposals have a particular kind of open-endedness. The notion of concern and imagination is open to indefinite modification in the light of research programmes in human biology, and also in the light of any research programmes that might be developed to improve our understanding of the consequences of any more specific ethical proposals which we might for the time-being include in our active set.

11.4 Ethical progress

The general principles we have now reached might be summed up by saying that, rather than accepting any theory as self-evident and any proposals as right, we need to examine their possible consequences and to keep them under review. If the OR community decided to push for clarity in the area of ethical proposals, it would mean joining in attempts to generate proposals for the modification of social institutions and in attempts to analyse the consequences of such modifications.

It might also mean joining in attempts to analyse the consequences of programmes aimed at modifying the total symbol–image–emotion field of individual people: for example, training, advertising, and the generation of military zeal. At the next higher level it might mean joining in attempts to analyse the consequences of programmes aimed at monitoring and controlling programmes aimed at modifying the total symbol–image–emotion field of individual people: for example, advertising control or training inspection.

Which is perhaps a large enough field to avoid anyone feeling cramped, but which requires skills and interests which reach rather beyond the present range if OR communities, and which suggests new classes of programme in which to intervene.

12. Professional competence

In this final chapter I present a picture of articulate reflection before action and then use it in a brief discussion of what can be said about individual competence and organizational competence.

12.1 A pictorial model of articulate reflection before action

Like any model, figure 8 has its weakness as a representation of what it points to. Its value is that it provides an easily remembered form for reminding us of the conjectural status of reflection before action, that it can serve as an heuristic tool for prompting a consideration of what might not have otherwise been considered, that it provides a module that can be used suggestively at various levels of aggregation, and that at a higher level of aggregation it suggests views on the structure of relationships between various roles.

The model is drawn to act as a direct reminder that in an episode of articulate reflection:

(1) Active theories are drawn from an indefinitely large set of latent theories,

(2) Both active and latent theories depend on an indefinitely large cascade of supporting theories, indefinitely large in extent behind each sector of the supporting cascade, reaching back indefinitely far into attempts at concept clarification and theory checking, and remaining for the most part unexplored by virtue of conscious decision or by virtue of resting on the unexamined use of natural language,

(3) Articulate reflection includes the acts of including and excluding theories from the active set,

(4) Active proposals are drawn from an indefinitely large set of latent proposals, including proposals about conditional action in various circumstances by

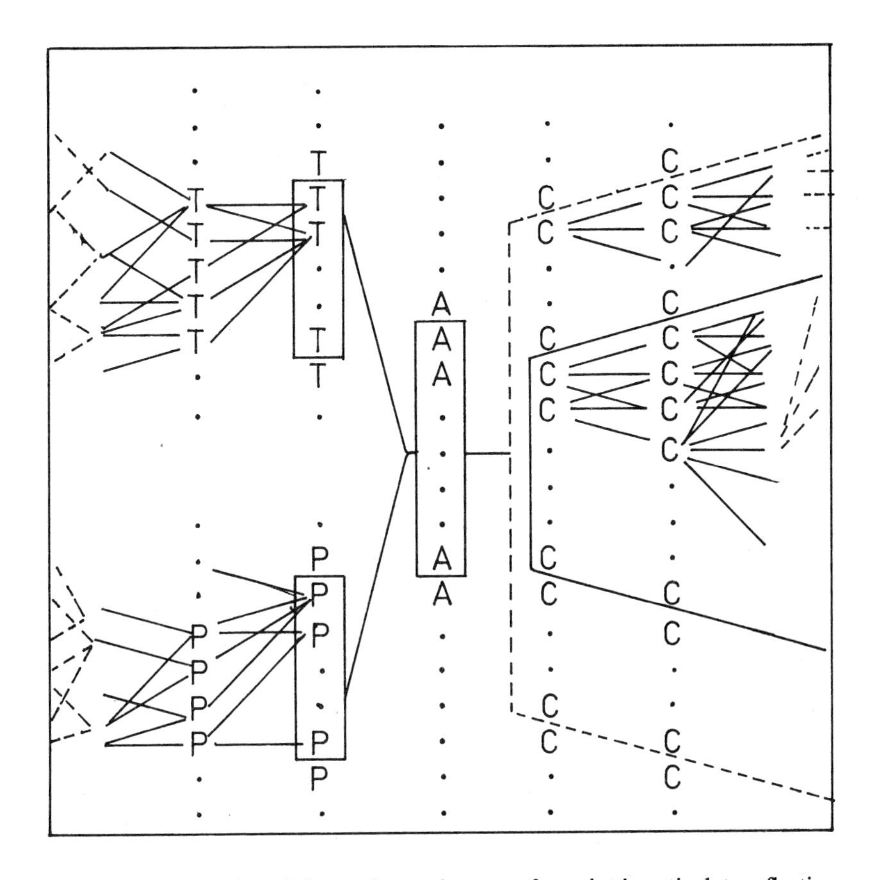

FIGURE 8 A reminder of the conjectural status of a point in articulate reflection before action. Numerous latent theories (T) supply selected active theories (boxed T); likewise numerous latent proposals (P) lead to active proposals (boxed P). Implemented actions (boxed A), based on the active theories and proposals, lead to cascades of consequences (C) proliferating into the future.

all the relevant action programmes and including proposals about measures of achievement,

(5) Both active and latent proposals presume an indefinitely large cascade of supporting proposals, indefinitely large behind each sector of the supporting cascade, and reaching back indefinitely into the traditions of society and into the conflicting views about them,

(6) Articulate reflection includes the acts of including and excluding proposals from the active set,

(7) Actively considered actions are drawn from an indefinitely large repertoire of imagined actions,

(8) Articulate reflection includes the act of including and excluding imagined actions from the active set,

(9) Articulate reflection includes the combining of theories and proposals to conjecture the consequences of imagined actions, including the consequences conditional on future circumstances and including estimates of proposed measures of achievement,

(10) Actions lead to an indefinitely large cascade of consequences, indefinitely large in extent at any point in time and extending indefinitely far into the future,

(11) Conceptualization of, or attention to, relevant consequences is incomplete: outside what is actively considered is a wider set which would be relevant to the theorizing and proposing, if the reflectors did but know it, and beyond that and only fuzzily distinguishable is the wider set of all imaginable consequences,

(12) Articulate reflection includes the acts of including and excluding conceptualizable consequences from the actively considered set of consequences.

A number of ideas do not find direct expression in the model. For example, it does not remind us that an episode of articulate reflection can be conducted in natural language, or in some mix of natural and formal language. Nor does it remind us that the generation of new ideas is an important heuristic consequence of reflection. Nor does it represent any form of development through time of either the reflection or of the reality which is its context. That is, it is a model of a point in the process of reflection: a sort of snapshot view.

12.2 Levels of aggregation

It seems to me that the picture emphasizes the technological and engineering character of articulate intervention, and that has led me on occasion to use the model to discuss the nature of science and engineering with various professional groups.

With the picture of figure 8, an engineer is concerned with being aware of the latent content of four sets and with selecting and using four active sets under the proposal that he plans to achieve some consequences, whether it be something as simple as the design of a small component or as complex as the construction of a highly automated factory.

In contrast, a pure scientist would seem to be more concerned with the proposal that he critically investigate the standing of a particular theory and its supporting cascade, or that he undertake action intended heuristically to suggest relevant new theory. That is, it would seem that the role of scientist does not include responsibility for the assembly of active sets. Of course, there are many people who are called scientists who combine an engineering and a scientific role and whose product is design. That could be said to be true of would-be-scientific intervention into the organizational goings-on of a programme as well as the design of physical products.

So, for example, a design engineer for a power station will have in mind the main proposals for the type of power station required, perhaps down to extensive and tight technical specifications which define a set of consequences to achieve. He will be aware of extensive proposals on codes of practice. He will be expected to assemble suitable sets of active theories about suitable sets of possible actions to achieve the stated consequences. At the same time he will be expected to assume responsibility for ensuring that a large and indefinite set of other consequences do not violate a wide and undefined set of latent proposals about them. So for example he will be expected to produce a design which leads to the required level of electricity production, but in addition it will be expected that at no future time should the equipment prove to have been designed in ways which make it harder to deal with than other equipment of that generation, for any kind of minor or major repair that actually arises. However, the designer may well not be reproached if it is unable to withstand heavily armed riot, or if it is not usable by the technically unsophisticated remnant of a major nuclear war, or if it does not later lend itself immediately to major differences in organization suggested by new political ideas.

Analogous achievements are required of would-be-scientific interveners who agree to leave behind some permanent alterations in the conduct of deliberative argument. Not only is it expected that the overt proposed consequences will be achieved but also that there will be no surprises that can thereupon be declared to be a regression from what might have been expected from a competent intervener.

It is possible to use the picture of articulate reflection at other levels of aggregation than the one described. At a much lower level of aggregation it could be used as a framework for describing the reflection that accompanies the making of a cake. At a much higher level of aggregation it could be used as a framework to suggest the roles of different groups: for example, the engineers who specify what a power station is to achieve are the proposers, the engineers who prepare designs to meet the specification are the theorizers, the engineers who construct the station are actors, and the power station and its ongoing organization and the slightly warmer surroundings are some of the consequences.

But it is the first level of aggregation I described that is of interest here, since it makes possible a simple analysis of professional competence.

12.3 A view of professional competence

The picture of articulate reflection suggests the following notion of what we might mean by *competence*:

> Competence is being relatively successful in achieving certain *closed* classes of consequences,
> from a repertoire of actions which is added to from time to time,
> by selecting and using active theory sets from specific classes of theory,
> and by selecting and using active proposal sets from specific classes of proposal.

This notion of competence will serve for all levels of skill, from straightforward repetitive jobs, through ordinary professional groupings, to specialists with almost unique skills. *Negligence* is then a matter of the assessment by some important reference group that there has been a significant and avoidable omission in the consequences considered, the actions considered and taken, the theories known or selected, or the proposals known and responded to.

If we want the ideas to apply particularly to professional competence we would have to add that by professional we meant that:

> There are no other groups of people who are consistently better over that set of activities,
> Some of the theories, proposals, and repertoire of actions are not generally available in the population and represent a time-consuming and coherent achievement.
> The achieving of the consequences is undertaken for sponsoring programmes.

12.4 Efficient professional competence

It is not necessarily the case that a professional group is yet at all close to what it might in principle be able to achieve. To consider this we might start by supposing that there are already explicit proposals about a set of consequences to be achieved and that there are explicit, latent or implicit proposals about what measures of achievement are to be used, including any notion of tolerance in not achieving the precise set of consequences exactly. If we work backwards from the desired consequences, there will in principle be a large set of alternative programmes of professional activity by which the consequences can be tolerably well achieved. That set of alternative programmes, many of which may not yet have been imagined by anyone, will constitute a set of *sufficient* programmes.

From that large imagined set of programmes there will be a smaller set of programmes, each of which on the measures of achievement to be used is entirely better than one or more of the programmes to be excluded, and none of which is entirely better than any other programme to be included. That imaginary smaller set constitutes a set of *efficient* programmes: each different bundle of measures of achievement may imply a different set of efficient programmes.

I need hardly say that for most purposes we have little idea of what is actually sufficient, because our understanding is conjectural. Nor do we have much idea of what is efficient for a particular person, because our imperfect understanding of his present and future bundle of evaluation proposals is compounded with our conjectural understanding. So the idea of an efficient programme for anyone is only an ideal towards which we can make hesitant steps.

Yet progress towards efficient programmes is the aim of articulate intervention. It is what has characterized human achievement over the long term. It is what gives a finite human population infinite potential. It is what we note has happened when the advances of one generation are the commonplaces of the next. And it is what I hope I have helped you with, in writing this book!

References

Ackoff, R. L. *Scientific Method; Optimizing Applied Research Decisions*, Wiley, 1962.
Ackoff, R. L. and Emery, F. E., *On Purposeful Systems*, Tavistock, 1972.
Bellman, R., Dynamic programming: a reluctant theory. In *New Methods of Thought and Procedure* (ed. F. Zwicky and A. G. Wilson), Springer, 1967.
Boothroyd, H., *On the theory of operational research*, CIEBR Paper 51, University of Warwick, 1974.
Boothroyd, H., Describing operational research. In *Operational Research '75* (ed. K. B. Haley), North-Holland, 1976.
Churchman, C. W., *Prediction and Optimal Decision*, Prentice-Hall, 1961.
Churchman, C. W., *The Design of Enquiring Systems*, Basic, 1972.
Edie, L. C., Traffic delays at toll booths, *Opns Res.*, **2,** 107–138, 1954.
Festinger, L., *A Theory of Cognitive Dissonance*, Row-Peterson, 1957.
Feyerabend, P. K., *Against Method*, NLB, 1975.
Flagle, C. D., Huggins, W. H. and Roy, R. H., *Operations Research and Systems Engineering*, John Hopkins, 1964.
Foucault, M., *The Archaeology of Knowledge*, Tavistock, 1972.
Friend, J. K. and Jessop, W. N., *Local Government and Strategic Choice*, Tavistock, 1969.
Harrison, P. J. and Stevens, C. F., Bayesian forecasting, *J. R. Statist. Soc. B*, **38,** 205–247, 1976.
Hillier, F. S. and Lieberman, G. J., *Introduction to Operations Research*, Holden-Day, 1974.
Howard, N., *Paradoxes of Rationality; Theory of Metagames and Political Behaviour*, M.I.T., 1971.
Kuhn, T. S., *The Structure of Scientific Revolutions*, University of Chicago, 1962.

Lakatos, I., Changes in the problem of inductive logic. In *The Problem of Inductive Logic* (ed. I. Lakatos), North-Holland, 1968.

Lakatos, I., Falsification and the methodology of scientific research programmes. In *Criticism and the Growth of Knowledge* (ed. I. Lakatos and A. Musgrave), Cambridge, 1970.

Magee, B., *Popper*, Fontana Collins, 1973.

McClone, R. R., *The Training of Mathematicians; A Research Report*, SSRC, 1973.

McCloskey, J. F. and Trefethen, F. N., eds., *Operations Research for Management, Volume I*, John Hopkins, 1954.

McCloskey, J. F. and Coppinger, J. M., eds., *Operations Research for Management, Volume II*, John Hopkins, 1956.

Oakeshott, M., *On Human Conduct*, Clarendon, 1975.

Phillips, D. L., *Abandoning Method; Sociological Studies in Methodology*, Jossey-Bass, 1973.

Popper, K. R., *The Logic of Scientific Discovery*, Hutchinson, 1959.

Popper, K. R., *The Poverty of Historicism*, Routledge, 1957.

Popper, K. R., *The Open Society and its Enemies*, Routledge, 1945.

Popper, K. R., *Conjectures and Refutations*, Routledge, 1965.

Popper, K. R., *Objective Knowledge*, Oxford, 1972.

Silverman, D., *The Theory of Organisations; A Sociological Framework*, Heinemann, 1970.

Waddington, C. H., *O.R. in World War II*, Elek, 1973.

Wagner, H. M., *Statistical Management of Inventory Systems*, Wiley, 1962.

Wagner, H. M., *Principles of Operations Research*, Prentice-Hall, 1969.

White, D. J., *Decision Theory*, Allen and Unwin, 1969.

White, D. J., *Decision Methodology*, Wiley, 1975.

Guidelines for the practice of operations research, *Opns Res*, **19,** 1123–1258, 1971, with rejoinders in *Mgmt Sci.*, **18,** B608–B629, 1972.

Priestley, J. B., *Anyone for Tennis?*, unpublished television play, 1968.

Science in War, Penguin, 1940.

Tocher, K. D., remark in Operational Research Society Annual Conference, 1976.

Note: From the beginning of 1978 *Operational Research Quarterly* was re-titled *Journal of the Operational Research Society* and published monthly.

Author Index

Subject Index